D0354531

Living and Learning with New Media

This report was made possible by grants from the John D. and Catherine T. MacArthur Foundation in connection with its grant making initiative on Digital Media and Learning. For more information on the initiative visit www.macfound.org.

The John D. and Catherine T. MacArthur Foundation Reports on Digital Media and Learning

The Future of Learning Institutions in a Digital Age by Cathy N. Davidson and David Theo Goldberg with the assistance of Zoë Marie Jones

Living and Learning with New Media: Summary of Findings from the Digital Youth Project by Mizuko Ito, Heather Horst, Matteo Bittanti, danah boyd, Becky Herr-Stephenson, Patricia G. Lange, C. J. Pascoe, and Laura Robinson with Sonja Baumer, Rachel Cody, Dilan Mahendran, Katynka Z. Martínez, Dan Perkel, Christo Sims, and Lisa Tripp

Confronting the Challenges of Participatory Culture: Media Education for the 21st Century by Henry Jenkins (P.I.) with Ravi Purushotma, Margaret Weigel, Katie Clinton, and Alice J. Robison

The Civic Potential of Video Games by Joseph Kahne, Ellen Middaugh, and Chris Evans

Young People, Ethics, and the New Digital Media: A Synthesis from the Good-Play Project by Carrie James with Katie Davis, Andrea Flores, John M. Francis, Lindsay Pettingill, Margaret Rundle, and Howard Gardner

Living and Learning with New Media

Summary of Findings from the Digital Youth Project

Mizuko Ito, Heather Horst, Matteo Bittanti, danah boyd, Becky Herr-Stephenson, Patricia G. Lange, C. J. Pascoe, and Laura Robinson

with Sonja Baumer, Rachel Cody, Dilan Mahendran, Katynka Z. Martínez, Dan Perkel, Christo Sims, and Lisa Tripp

The MIT Press
Cambridge, Massachusetts
London, England

For information about special quantity discounts, please email special_sales@mitpress.mit.edu.

This book was set in Stone Serif and Stone Sans by the MIT Press. Printed and bound in the United States of America.

Library of Congress Cataloging-in-Publication Data

Living and learning with new media : summary of findings from the digital youth project / Mizuko Ito ... [et al.].
    p. cm.—(The John D. and Catherine T. MacArthur Foundation reports on digital media and learning)
Includes bibliographical references.
ISBN 978-0-262-51365-4 (pbk. : alk paper)
1. Mass media and youth—United States. 2. Digital media—Social aspects—United States. 3. Technology and youth—United States.
4. Learning—Social aspects. I. Ito, Mizuko.
HQ799.2.M352U655 2009   302.23'108350973—dc22   2009008614

10 9 8 7 6 5 4 3 2 1

Contents

Series Foreword

The John D. and Catherine T. MacArthur Foundation Reports on Digital Media and Learning, published by the MIT Press, present findings from current research on how young people learn, play, socialize, and participate in civic life. The Reports result from research projects funded by the MacArthur Foundation as part of its $50 million initiative in digital media and learning. They are published openly online (as well as in print) in order to support broad dissemination and to stimulate further research in the field.

Acknowledgments

The research for this project was funded by a grant from the John D. and Catherine T. MacArthur Foundation and the project was housed at the Institute for Multimedia Literacy and the Annenberg Center for Communication at the University of Southern California and at The Institute for the Study of Social Change and the School of Information at the University of California, Berkeley. We would especially like to thank Julia M. Stasch and Constance M. Yowell at the MacArthur Foundation, and the coprincipal investigators on this project, Michael Carter, Peter Lyman, and Barrie Thorne, for their guidance throughout. This project was also guided and supported by Diane Harley and the following administrative and technical staff: Josie Acosta, Steve Adcook, Chris Badua, Kathleen Kuhlmann, Shalia MacDonald, Mariko Oda, Willy Paredes, Janice Tanigawa, Chris Wittenberg, and Evelyn Wong.

In addition to the authors and contributors to this report, we had many research assistants and collaborators who enriched this project along the way. Max Besbris, Brendan Callum, Allison Dusine, Sam Jackson, Lou-Anthony Limon, Renee Saito, Judy Suwatanapongched, and Tammy Zhu were research assistants and vital informants and experts in all things digital and youth. We also benefited from working with our collaborators on this project, Natalie Boero, Carrie Burgener, Scott Carter, Juan Devis, Paul Poling, Nick Reid, Rachel Strickland, and Jennifer Urban. The Berkman Center for Internet and Society at Harvard, the Pew Internet & American Life Project, and the Annenberg Foundation also were institutional collaborators in this research.

Our thinking for this report was enriched by reviews of our research from John Seely Brown, Paul Duguid, Jabari Mahiri, Daniel Miller, Katie Salen, Ellen Seiter, and Barry Wellman. We are grateful to Karen Bleske for her careful editing and to Eric Olive for being our Web guru. Throughout this project, we were blessed with many thoughtful colleagues who attended our project meetings and provided guidance along the way: Sasha Barab, Brigid Barron, Suzy Beemer, Linda Burch, Lynn Schofield Clark, Michael Cole, Brinda Dalal, Dale Dougherty, Penelope Eckert, Nicole Ellison, James Paul Gee, David Goldberg, Shelley Goldman, Joyce Hakansson, Eszter Hargittai, Glynda Hull, Lynn Jamieson, Henry Jenkins, Joseph Kahne, Amanda Lenhart, Jane McGonigal, Ellen Middaugh, Kenny Miller, Alesia Montgomery, Kimiko Nishimura, John Palfrey, Nichole Pinkard, Alice Robison, Ryan Shaw, Lissa Soep, Reed Stevens, Deborah Stipek, Benjamin Stokes, Pierre Tchetgen, Doug Thomas, Avril Thorne, and Margaret Weigel.

Finally, we would like to thank the many individuals, families, organizations, and online communities that welcomed us into their midst and educated us about their lives with new media. Although we cannot name all the individuals who participated in our study, we would like to express our gratitude to those whom we can name who facilitated our access to various sites and who acted as key "local" experts: Vicki O'Day for introducing Heather to Silicon Valley families; Tim Park, Carlo Pichay, and zalas for being Mizuko's *senpai* in the anime fandom; Enki, Wurlpin, and all of KirinTheDestroyers for taking Rachel under their wing; Tom Anderson, who helped danah get access to MySpace; the people of YouTubia who spoke with Patricia and shared their videos; and all of the youth media, middle-school, and high-school educators who opened their doors to us.

Executive Summary

Young people in the United States today are growing up in a media ecology where digital and networked media play an increasingly central role. Even youth who do not possess computers and Internet access at home are participants in a shared culture where new social media,[1] online media distribution, and digital media production are commonplace among their peers and in their everyday school contexts. The implications of this *new media ecology* weigh heavily on the minds of parents and educators alike, who worry about the changes new media may present for learning and literacy as well as for the process of growing up in American society.

This report summarizes the results of a three-year ethnographic study, funded by the John D. and Catherine T. MacArthur Foundation, examining young people's participation in the new media ecology. It represents a condensed version of a longer treatment of the project findings (Ito et al., forthcoming). We present empirical data of new media in the lives of

American youth in order to reflect on the relationship between new media and learning. In our research, one of the largest qualitative and ethnographic studies of American youth culture, we examine what sociality among young people actually looks like in this new media ecology as well as how the emergence of *networked public culture* may shape and transform social interaction, *peer-based learning,* and *new media literacy* among young people.

This research was designed to address a gap in the literature surrounding the role of digital media in the lives of American youth. While there are a growing number of quantitative studies surveying the overall distribution of youth digital media practices, most qualitative research is based on single case studies, making it difficult to document the broader social and cultural contours, as well as the overall diversity, in youth engagement with digital media. Given the lack of research in this area, our study was motivated by two primary research questions:

• How are new media being taken up by youth practices and agendas?
• How do these practices change the dynamics of youth-adult negotiations over literacy, learning, and authoritative knowledge?

In framing the analysis of this research, we believe that there are four key concepts that characterize the ways youth live and learn with new media and, in turn, our perspective on the practices and conditions that define young people's engagements with new media.

New Media Ecology   We use the term *new media* to describe a media ecology where more traditional media such as books, television, and radio are intersecting with digital media, specifically interactive media, online networks, and media for social communication. We use the metaphor of *ecology* to emphasize that the everyday practices of youth, existing structural conditions, infrastructures of place, and technologies are all dynamically interrelated; the meanings, uses, functions, flows, and interconnections in young people's everyday lives located in particular settings are also situated within young people's wider media ecologies.

Networked Publics   The term *networked publics* describes participation in public culture (Appadurai and Breckenridge 1988) that is supported by Internet and mobile networks. The growing availability of digital media-production tools and infrastructure, combined with the traffic in media across social connections and networks, is creating convergence between mass media and online communication (Benkler 2006; Ito 2009; Jenkins 2006; Shirky 2008; Varnelis 2009). Rather than conceptualize everyday media engagement as "consumption" by "audiences," the term *networked publics* foregrounds the active participation of a distributed social network in the production and circulation of culture and knowledge.

Peer-Based Learning Our attention to youth perspectives, as well as the high level of youth engagement in social and recreational activities online, determined our focus on the more informal and loosely organized contexts of peer-based learning. Our focus is on describing learning outside of school, primarily in settings of peer-based interaction. While adults often view the influence of peers negatively, as characterized by the term *peer pressure*, we approach these informal spaces for peer interactions as spaces of opportunity for learning.

New Media Literacy We examine the current practices of youth and query what kinds of literacies and social competencies they are defining as a particular generational cohort, experimenting with a new set of media technologies. To inform current debates over the definition of new media literacy, we describe the forms of competencies, skills, and literacy practices that youth are developing through media production and online communication in order to inform these broader debates.

Alongside the conceptual framework that structured our study, throughout this report we frame youth engagements with new media in terms of emerging practices, or *genres of participation*. This framework does not rely solely on distinctions based on given categories such as gender, class, or ethnic identity. Rather, we identified distinct, but interrelated, genres based on what we saw in our ethnographic material on youth practice and culture. Genres of participation help us interpret how media intersect with learning and participation. The first two genres focus upon the activities and perspectives that *motivate*, or drive, young people's use of new media.

Friendship-Driven Genres of Participation A friendship-driven genre of participation characterizes the dominant and mainstream practices of youth as they go about their day-to-day negotiations with friends and peers in given, local contexts that center on relationships fostered in school and other local community institutions.

Interest-Driven Genres of Participation An interest-driven genre of participation characterizes engagement with specialized activities, interests, or niche and marginalized identities. In contrast to friendship-driven participation, kids establish relationships that center on their interests, hobbies, and career aspirations rather than friendship per se.

In addition to the broad distinctions between friendship-driven and interest-driven genres of participation, we have identified three genres that correspond to differing *levels of commitment and intensity* in new media practices.

• *Hanging out* is primarily a friendship-driven genre of participation in which young people spend their casual social time with one another. In interest-driven groups that result in friendships, we also see hanging out activity, but most youth hanging out is with local friendship-driven networks. Sites such as MySpace and Facebook, and communications technologies such as instant messaging (IM) and text messaging, provide a lightweight means for youth to stay in ongoing social contact and to arrange real-life gatherings. Furthermore, new media provide a topic for conversation, in the form of forwarding and linking to interesting pieces of online media, as well as a focus for activity, such as when youth play social games together or share music. As we will illustrate, hanging out may also take place within the context of home and family life.

• *Messing around* represents the beginning of a more intense media-centric form of engagement. When messing around, young people begin to take an interest in and focus on the workings and content of the technology and media themselves, tinkering, exploring, and extending their understanding. Some activities that we identify as messing around including looking around and searching for information online as well as experimentation and play using a range of media, such as digital and video cameras, music and photo editing software, and other new media. Messing around is often a transitional genre, in which kids move between hanging out and friendship-driven forms of participation to more interest-driven genres of participation.

• *Geeking out* involves the more expertise-centered forms of interest-driven participation surrounding new media that we found among some of the gamers, fans, and media producers we encountered in our study. Geeking out involves intensive and frequent use of new and, at times, relatively obscure media, high levels of specialized knowledge, alternative models of status and credibility, and a willingness to bend and/or break social and technological rules.

Our practice-focused analysis of young people in the new media ecology enabled the documentation of the everyday lives of youth in the United States. It also structured the development of an empirically based paradigm for understanding learning and participation in contemporary networked publics. From this work, we suggest the following implications of our findings for the dynamics of youth-adult negotiations over literacy, learning, and authoritative knowledge:

Robust participation in networked publics requires a social, cultural, and technical ecology grounded in social and recreational practices.

Ongoing, lightweight, and relatively unrestricted access to digital-production tools and the Internet was a precondition for participation in most of the networked public spaces that are the focus of attention for U.S. teens. Further, much of this engagement is centered on access to social and commercial entertainment content that is generally frowned upon in formal educational settings.

Networked publics provide a context for youth to develop social norms in the context of public participation.

Networked publics have altered many of the conditions of hanging out and publicity for youth, even as they build on existing youth practices of socializing, flirting, and pursuing hobbies and interests. Contrary to fears that social norms are eroding online, we saw almost no evidence that participation in networked publics resulted in riskier behavior than teens engaged in offline, and their online communication is conducted in a context of public scrutiny and structured by well-developed norms of social appropriateness, a sense of reciprocity, and collective ethics.

Youth are developing new forms of media literacy that are keyed to new media and youth-centered social and cultural worlds.

Youth are developing a wide range of new literacy forms through their informal new media practices, including deliberately casual forms of online speech, formats for displaying public connections, and new forms of appropriative literacies such as customizing MySpace profiles, mashups, and remixes. Efforts to address new media literacy need to take into account the specific social and cultural contexts that are meaningful to youth.

Peer-based learning has unique properties that drive engagement in ways that differ fundamentally from formal instruction.

In both the friendship-driven and interest-driven sides peers help to drive learning. Peer-based learning is characterized by a context of reciprocity, in which participants believe they can both produce and evaluate knowledge and culture, and in which they can develop reputation and receive recognition from respected peers. In these settings, the focus of learning and engagement is not defined by institutional accountabilities but rather emerges from kids' interests and everyday social communication.

Living and Learning with New Media

Living and Learning with New Media: Summary of Findings from the Digital Youth Project

Digital media and online communication have become a pervasive part of the everyday lives of youth in the United States. Social network sites, online games, video-sharing sites, and gadgets such as iPods and mobile phones are now well-established fixtures of youth culture; it can be hard to believe that just a decade ago these technologies were barely present in the lives of U.S. children and teens. Today's youth may be engaging in negotiations over developing knowledge and identity, coming of age, and struggling for autonomy as did their predecessors, but they are doing this while the contexts for communication, friendship, play, and self-expression are being reconfigured through their engagement with new media. We are wary of the claims that there is a digital generation that overthrows culture and knowledge as we know it and that its members' practices are radically different from older generations' new media engagements. At the same time, we also believe that current youth adoption of digital media production and social media are occurring in a unique historical moment, tied to long-term and systemic changes in sociability and culture. While the pace of technological change may seem dizzying, the underlying practices of sociability, learning, play, and self-expression are undergoing a slower evolution, growing out of resilient social structural conditions and cultural categories that youth inhabit in diverse ways in their everyday lives. The goal of the digital youth study was to document a point in this changing ecology by looking carefully at how both the commonalities and diversity in youth new media practice are part of a broader social and cultural ecology.

Our values and norms surrounding education, literacy, and public participation are being challenged by a shifting landscape of media and communications where youth are central actors. Although questions about "kids these days" have a familiar ring to them, the contemporary version is somewhat unusual in how strongly it equates generational identity with technology identity, an equation that is reinforced by telecommunications and digital media corporations that hope to capitalize on this close identification. There is a growing public discourse (both hopeful and fearful) declaring that young people's use of digital media and communication technologies defines a generational identity distinct from that of their elders. In addition to this generational divide, these new-technology practices are also tied to what David Buckingham (2007, 96) has described as a "'digital divide' between in-school and out-of-school use." He sees this as "symptomatic of a much broader phenomenon—a widening gap between children's everyday 'life worlds' outside of school and the emphases of many educational systems." Both the generational divide and the divide between in-school and out-of-school learning are part of a resilient set of questions about adult authority in the education and socialization of youth. The discourse of digital generations and digital youth posits that new media empower youth to challenge the social norms and educational agendas of their elders in unique ways. This report, and the corresponding book (see Ito et al., forthcoming), questions and investigates these claims. How are new media being taken up by youth practices and agendas? And how do these practices change the dynamics of youth-adult negotiations over literacy, learning, and authoritative knowledge?

Despite the widespread assumption that new media are tied to fundamental changes in how young people are engaging with culture and knowledge, there is still relatively little research that investigates how these dynamics operate on the ground. This report summarizes a three-year ethnographic investigation of youth new media practices that aims to develop a grounded, qualitative evidence base to inform current debates over the future of learning and education in the digital age. Funded by the John D. and Catherine T. MacArthur Foundation as part of a broader initiative on digital media and learning, the study represents a $3.3 million investment to contribute to basic knowledge in this emerging area of research. The project began in early 2005 and was completed in the summer of 2008, with the bulk of fieldwork taking place in 2006 and 2007. This report represents a summary of a book reporting on the findings from this project, titled *Hanging Out, Messing Around, and Geeking Out: Kids Living and Learning with New Media*. This effort is unique among qualitative studies in the field in the breadth of the research and the number of case studies that it encompasses. Spanning 23 different case studies conducted by 28 researchers and collaborators, this study sampled from a wide range of different youth practices, populations, and online sites, all centered on the United States. We drew from 20 of these case studies to write our collaborative book. This study has a broad descriptive goal of documenting youth practices of engagement with new media and a more targeted goal of analyzing how these practices are part of negotiations between adults and youth over learning and literacy.

Research Approach

Although a growing volume of research is examining youth new media practice, we are still at the early stages of piecing together a more holistic picture of the role of new media in young people's everyday lives. A growing number of quantitative studies document the spread of new media and related practices among U.S. youth (Griffith and Fox 2007; Lenhart et al. 2007; Rainie 2008; Roberts, Foehr, and Rideout 2005). In addition to these quantitative indicators, there is a growing body of ethnographic case studies of youth engagement with specific kinds of new media practices and sites (some examples include Baron 2008; Buckingham 2008; Ito, Okabe, and Matsuda 2005; Ling 2004; Livingstone 2008; Mazzarella 2005). Although the United Kingdom has funded some large-scale qualitative studies on youth new media engagements (Holloway and Valentine 2003; Livingstone 2002), the United States has not had comparable qualitative studies that look across a range of different populations and new media practices. What is generally lacking in the literature overall, and in the United States in particular, is an understanding of how new media practices are embedded in a broader social and cultural ecology. While we have a picture of technology trends on one hand, and spotlights on specific youth populations and practices on the other, we need more work that brings these two pieces of the puzzle together. This study addresses this gap, through a large-scale ethnographic study that looks across a wide range of different youth populations and their new media practices. We approached the descriptive goal of our study with a research approach that was defined by ethnographic method, a youth-centered focus, and the study of the changing new media ecology.

Ethnography

An ethnographic approach means that we work to understand how media and technology are meaningful to people in their everyday lives. We rely on qualitative methods of interviewing, observation, and interpretive analysis in an effort to understand patterns in culture and social practices from the point of view of participants themselves, rather than beginning with our own categories. The goal is to capture the youth cultures and practices related to new media, as well as the surrounding context— such as peer relations, family dynamics, local community institutions, and broader networks of technology and consumer culture. The strength of this approach is that it enables us to surface, from the empirical material, what the important categories and structures are that determine new media practices and learning outcomes. This approach does not lend itself to testing existing analytic categories or targeted hypotheses but rather to asking more fundamental questions about what the relevant factors and categories of analysis are. We believe that an initial broad-based ethnographic understanding, grounded in the actual contexts of behavior and local cultural understandings, is crucial in grasping the contours of a new set of cultural categories and practices.

Focus on Youth

Adults often view children in terms of developmental "ages and stages," focusing on what they will become rather than as complete beings "with ongoing lives, needs and desires" (Corsaro 1997, 8). By contrast, we take a sociology-of-youth-and-childhood approach, which means that we take youth seriously as actors in their own social worlds and look at childhood as a socially constructed, historically variable and contested category (Corsaro 1997; Fine 2004; James and Prout 1997; Wyness 2006). Our work has focused on documenting the everyday new media practices of youth in their middle-school and high-school years, and we have made our best effort at documenting the diversity of youth identity and practice. We have also engaged, to a lesser extent, with parents, educators, and young adults who participate in or are involved in structuring youth new media practices. Readers will see the subjects of this research referred to by a variety of age-related names. We use the term "kids" for those 13 and under, "teens" for those aged 13 to 18, and we use the term "young people" to refer to teens and young adults aged 13 to 30. We use the term "youth" to describe the general category of youth culture that is not clearly age demarcated but that centers on the late teenage years. Interviews with young adults are included to provide a sense of adult participation in youth practice as well as to provide retrospective accounts of growing older with new media. While age-based categories have defined our object of study, we are interested in documenting how these categories are historically and culturally specific, and how new media use is part of the redefinition of youth culture and "age-appropriate" forms of practice.

New Media

We use the term *new media* to describe a media ecology where more traditional media, such as books, television, and radio, are intersecting with digital media, specifically interactive media and media for social communication (Jenkins 2006). In contrast to work that attempts to isolate the specific affordances of digital production tools or online networks, we are interested in the convergent media ecology that youth inhabit today. We have used the term *new media* rather than terms such as *digital media* or *interactive media* because the moniker of "the new" seemed appropriately situational, relational, and protean, and not tied to a specific media platform. Our work has focused on those practices that are "new" at this moment and that are most clearly associated with youth culture and voice, such as engagement with social network sites, media fandom, and gaming. The aim of our study is to describe media engagements that are specific to the life circumstances of current youth, at a moment when we are seeing a transition to participation in digital media production and networked publics. Following from our youth-centered approach, the new media practices we examine are almost all situated in the social and recreational activities of youth rather than in contexts of explicit instruction.

The Study

The Digital Youth Project was led by four principal investigators, Peter Lyman, Mizuko Ito, Michael Carter, and Barrie Thorne. During the course of the three-year research grant (2005–2008), seven postdoctoral researchers,[1] six doctoral students,[2] nine MA students,[3] one JD student,[4] one project assistant,[5] seven undergraduate students,[6] and four research collaborators[7] from a range of disciplines participated in and contributed fieldwork materials to the project. Our project was designed to document, from an ethnographic perspective, the learning and innovation that accompany young people's everyday engagements with new media in informal settings. Specifically, our focus on youth-centered practices of play, communication, and creative production located learning in contexts that are meaningful and formative for youth, including friendships and families, as well as young people's own aspirations, interests, and passions. The practices we focused on moved across geographic and media spaces and, as a result, our ethnography incorporated multiple sites and research methods, varying from questionnaires, surveys, semi-structured interviews, diary studies, observation, and content analyses of media sites, profiles, videos, and other materials. Collectively, we conducted 659 semi-structured interviews, 28 diary studies, and focus group interviews with 67 individuals. Interviews were conducted informally with at least 78 individuals and we also participated in more than 50 research-related events, such as conventions, summer camps, award ceremonies, or other local events. Complementing our interview-based strategy, we also carried out more than 5,194 observation hours, which were chronicled in regular field notes, and we have collected 10,468 profiles, posts from 15 online discussion group forums, and more than 389 videos as well as numerous materials from classroom and after-school contexts. Our Digital Kids Questionnaire was completed by 402 participants, with 363 responses from people under the age of 25. Our analysis for our joint book and report draws on work across 20 distinctive research projects that were framed in terms of four main areas: homes and neighborhoods, institutional spaces, online sites, and interest groups.[8]

Homes and Families

We focused on homes and families in urban, suburban, and rural contexts in order to understand how new media and technologies shape the contours of kids' home lives and, in turn, how different family structures and economic and social positions may structure young people's media ecologies (Bourdieu 1984; Holloway and Valentine 2003; Livingstone 2002; Silverstone and Hirsch 1992). Through the five projects outlined below, we focused on young people ranging from ages 8 to 20 with attention to developing an understanding of the influence of ethnic, racial, gender, class, and economic distinctions on young people's media and technology practices and participation (Chin 2001; Escobar 1994; Pascoe 2007; Seiter 2005; Thorne 2008).

Coming of Age in Silicon Valley Heather A. Horst examined the appropriation of new media and technology in Silicon Valley, California, among families with children ranging from the ages of 8 to 18. Horst's research focused on the gendered and generational dynamics of new media use in families. Throughout this report, research from this study will be referenced as *Silicon Valley Families*.

Digital Media in an Urban Landscape Lisa Tripp, Becky Herr-Stephenson, and Katynka Z. Martínez coupled participant observation in the classrooms of Los Angeles–based teachers involved in a professional-development program for media arts and technology with participant observation in after-school programs with home interviews, which were conducted in English and/or Spanish. Examples drawn from this study are noted as *Pico Union Families, Computer Club Kids, Animation around the Block, L.A. Youth and Their Community Center,* and *Los Angeles Middle Schools.*

Discovering the Social Context of Kids' Technology Use Dan Perkel and Sarita Yardi used digital-photography diary studies to develop an understanding of the technology practices of kids entering middle school. This project also developed methods for use of camera phones in diary studies, which supplemented the interviews and participation in Horst, Martínez, and C. J. Pascoe's research on new media in everyday life. Research drawn from this study is labeled *Digital Photo-Elicitation with Kids* throughout this report.

Freshquest Megan Finn, David Schlossberg, Judd Antin, and Paul Poling's study focused on the roles of media and technologies in the lives of teenagers through a study of technology-mediated communication habits of freshman students at the University of California, Berkeley. This study included interviews and surveys with Berkeley freshmen as well as a smaller sample of first-year students at a community college outside of San Francisco. We use the name *Freshquest* to indicate material that emerged from this study.

Living Digital: Teens' Social Worlds and New Media C. J. Pascoe and Christo Sims conducted a multi-sited ethnographic project in order to analyze how teenagers communicate, negotiate social networks, and craft a unique teen culture using new media. Whereas Pascoe carried out research in the San Francisco Bay Area, Sims interviewed teenagers in rural California and Brooklyn, New York. Examples from Pascoe's work are labeled *Living Digital*. Christo Sims's research materials are denoted as *Rural and Urban Youth*.

Learning Institutions: Media Literacy Programs and After-School Programs

In the past two decades, researchers interested in "informal learning" have increasingly turned their attention to institutions, such as libraries, after-school programs, and museums, that structure learning experiences that differ from those in school (see Barron 2006; Bekerman, Burbules, and Silberman-Keller 2006). As institutions temporally and spatially situated between the dominant institutions in kids' lives—school and family—after-school programs and spaces offered potential for observing instances of informal learning, particularly given the increasing importance of after-school and enrichment programs in American public education. In light of the possibilities of these spaces, four of our research projects focused on media literacy and after-school programs in an effort to understand how they fit into the lives of young people.

Information the Wiki Way Laura Robinson examined the role played by material resources in everyday information-seeking contexts among economically disadvantaged youth at a high school in an agricultural region of Central California. Research materials from this study are noted as *Wikipedia and Information Evaluation*.

Teaching and Learning with Multimedia Lisa Tripp and Becky Herr-Stephenson explored the complex relationships between multimedia production projects undertaken in middle-school classrooms and students' out-of-school experiences with multimedia, with particular attention to the gaps and overlaps of media use within the contexts of home and school. We use the phrase *Los Angeles Middle Schools* to identify material from this study. The material from the *Computer Club Kids, Animation around the Block,* and *L.A. Youth and their Community Center* projects carried out by Katynka Z. Martínez was also integrated within this analytic framework.

The Social Dynamics of Media Production in an After-School Setting Judd Antin, Dan Perkel, and Christo Sims looked at how the students from low-income neighborhoods negotiate and appropriate the structured and unstructured aspects of after-school programs in order to learn new technical skills, socialize with new groups of friends, and take advantage of the unique access to both technical and social resources that are often lacking in their homes and schools. We cite material drawn from this study with the label *The Social Dynamics of Media Production*.

Networked Sites

Six of our research projects concentrated on understanding emerging practices surrounding membership and participation on a series of Web sites that dominated young people's media ecologies between 2005 and 2007. Rather than restricting our focus to bounded spaces or locales (Appadurai 1996; Basch, Schiller, and Szanton-Blanc 1994; Gupta and Ferguson 1997), we started our research by focusing on these popular sites, concentrating our efforts on understanding practices as they spanned online and offline settings without privileging one context as more or less authentic or more or less virtual (boyd 2007; Kendall 2002; Lange 2008; Miller and Slater 2000).

Broadcast Yourself: Self-Production through Online Video-Sharing on YouTube   Sonja Baumer's study analyzed self-production as an agentive act that expresses the fluidity of identity achieved through forms of semiotic action and through practices such as self-presentation, differentiation and integration, self-evaluation, and cultural commentary. We use the short title *Self-Production through YouTube* to indicate research material from this study.

Life in the Linkshell: The Everyday Activity of a Final Fantasy Community   Rachel Cody looked at the massively multiplayer online role-playing game *Final Fantasy XI* in order to understand how social activity extends beyond the game into Web sites, message boards, and instant-messenger programs, and encourages a level of collaboration that is impossible within the game. Research material from this study is noted as *Final Fantasy XI.*

Teen Sociality in Networked Publics   danah boyd examined the ways in which teens use sites such as MySpace and Facebook to negotiate identity, socialize with friends, and make sense of the world around them. We cite material from this study with the label *Teen Sociality in Networked Publics.*

Thanks for Watching: A Study of Video-Sharing Practices on You-Tube   Patricia G. Lange focused on the ways in which YouTubers learn how to represent themselves and their work in order to become accepted members of groups who share similar media-based affinities through making videos and reacting to feedback. *YouTube and Video Bloggers* denotes material used from Lange's research.

The Practices of MySpace Profile Production   Dan Perkel investigated how young people create MySpace pages, with particular attention to the sociotechnical practices and infrastructure of profile making, such as social support and assistance, the location of visual and audio material online, and copying and pasting snippets of code. Material from Perkel's study is noted as *MySpace Profile Production.*

Virtual Playgrounds: An Ethnography of Neopets   Heather A. Horst and Laura Robinson explored cultural products and knowledge creation surrounding the online world of Neopets, with particular attention to how participants develop notions of reputation, expertise, and o.her forms of identification. We use the short title *Neopets* to refer to research from this study.

Interest-Based Communities

Recognizing the tremendous transformations in the empirical and theoretical work on youth subcultures, new media, and popular culture through recent decades (Cassell and Jenkins 1998; Gilroy 1987; Hall and Jefferson 1976; Hebdige 1979, 1987; Jenkins 1992; McRobbie and Garber 2000; Seiter 1993; Snow 1987), researchers across our project focused on the modes of expression, circulation, and mobilization of youth subcultural forms in and through new media. Six of our research projects focused on popular and subcultural forms and the changing relationships between producers, consumers, and participants through interest-driven engagements with new media.

Game Play Matteo Bittanti examined the complex relationship between teenagers and video games through a focus on the ways in which gamers create and experiment with different identities, learn through informal processes, craft peer groups, develop a variety of cognitive, social, and emotional skills, and produce significant textual artifacts through digital play. We use the label *Game Play* to indicate material derived from this study.

Hip-Hop Music and Meaning in the Digital Age Dilan Mahendran explored the practices of amateur music making in the background of hip-hop culture in San Francisco Bay Area afterschool settings. The study sought to understand the centrality of music listening and making by both enthusiasts and youth in general as world-disclosing practices that challenge the assumption that youth are simply passive consumers. All material from this study is indicated by the phrase *Hip-Hop Music Production*.

Mischief Managed   Rebecca Herr-Stephenson investigated multimedia production undertaken by young Harry Potter fans and the role technology plays in facilitating production and distribution of fan works. We note material drawn from this study with the label *Harry Potter Fandom*.

No Wannarexics Allowed   C. J. Pascoe and Natalie Boero focused on the construction of online eating-disorder communities through an analysis of pro-anorexia (ana) and pro-bulimia (mia) discussion groups, with particular attention to how the "ana" and "mia" lifestyles are produced and reproduced in these online spaces. All examples from this study are cited by the phrase *Pro-Eating Disorder Discussion Groups.*

Team Play: Kids in the Café   Arthur Law conducted ethnographic observations at a cybercafé where teens gathered to play online video games in collaborative team engagements, examining the construction of friendships and teamwork through networked gaming. We use the title *Team Play* to reference material from this study.

Transnational Anime Fandoms and Amateur Cultural Production   Mizuko Ito examined a highly distributed network of overseas fans of Japanese animation, focusing on how the fandom organized and communicated online and how it engaged in creative production through the transformative reuse of commercial media. We use the short title *Anime Fans* when referencing material from this study.

Conceptual Framework

Our work is guided by four key analytic foci that we apply to our ethnographic material: *genres of participation, networked publics, peer-based learning,* and *new media literacy.* In examining these different areas, we draw from existing theories that are part of the "social turn" in literacy studies, new media studies, learning theory, and childhood studies.

Genres of Participation

In order to understand new media engagement, we draw from models of learning that look to the learning in everyday activity and rely on a notion of social and cultural participation (Jenkins 1992, 2006; Karaganis 2007; Lave and Wenger 1991). We see learning with new media as a process of participation in shared culture and sociability as it is embodied and mediated by new technologies. In our descriptions of youth practice, we rely on a framework of "genres of participation" to describe different modes or conventions for engaging with new media (Ito 2003, 2008). Instead of looking to rigid categories that are defined by formal properties, genres of participation are a way of identifying, in an interpretive way, a set of social, cultural, and technological characteristics that are recognizable by participants as defining a set of practices.

We have not relied on distinctions based on given categories such as gender, class, or ethnic identity. Our genres are based on what we saw in our ethnographic material, patterns that helped us and our research participants interpret how media intersect with learning and participation. By describing these forms of participation as genres, we hope to avoid the assumption that these genres attach categorically to individuals. Rather, just as an individual may engage with multiple media genres, we find that youth will often engage in multiple genres of participation in ways that are situationally specific. We have also avoided categorizing practice based on technology- or media-centric parameters, such as media type or measures of frequency or media saturation. Genres of participation provide ways of identifying the sources of diversity in how youth engage with new media in a way that does not rely on a simple notion of "divides" or a ranking of more or less sophisticated media expertise. Instead, these genres represent different investments that youth make in particular forms of sociability and differing forms of identification with media genres.

▪ By *friendship-driven* genres of participation, we refer to the dominant and mainstream practices of youth as they go about their day-to-day negotiations with friends and peers. These friendship-driven practices center on peers whom youth encounter in the age-segregated contexts of school but might also include friends and peers whom they meet through religious groups, school sports, and other local activity groups. For most youth, these local friendship-driven networks are their primary source of affiliation, friendship, and romantic partners, and their lives online mirror this local network. MySpace and Facebook are the emblematic online sites for these sets of normative practices.

• In contrast to friendship-driven practices, with *interest-driven* genres of participation, specialized activities, interests, or niche and marginalized identities come first. Interest-driven practices are what youth describe as the domain of the geeks, freaks, musicians, artists, and dorks, who are identified as smart, different, or creative, and who generally exist at the margins of teen social worlds. Youth find a different network of peers and develop deep friendships through these interest-driven engagements, but in these cases the interests come first, and structure the peer network and friendships, rather than vice versa. These are contexts where youth find relationships that center on their interests, hobbies, and career aspirations. It is not about the given social relations that structure youth's school lives but about both focusing and expanding an individual's social circle based on interests. Although some interest-based activities such as sports and music have been supported through schools and overlap with young people's friendship-driven networks, other kinds of interests require more far-flung networks of affiliation and expertise.

Friendship-driven and interest-driven genres provide a broad framework for identifying what we saw as the most salient social and cultural distinction that differentiated youth new media practice. In addition, we have identified three genres of participation that describe different degrees of commitment to media engagement: *hanging out, messing around,* and *geeking out.* These three genres are a way of describing different levels of intensity and sophistication in relation to media engagement with reference to social and cultural context, rather than relying exclusively on measures of frequency or assuming that certain forms of media or technology automatically correlate with "high-end" and "low-end" forms of media literacy. In the second half of this report, we present an overview of our research findings in terms of these three genres of participation and related learning implications.

Participation in Networked Publics

We use the term *networked publics* to describe participation in public culture (Appadurai and Breckenridge 1988) that is supported by online networks. The growing availability of digital media-production tools, combined with online networks that traffic in rich media, is creating convergence between mass media and online communication (Benkler 2006; Ito 2008b; Jenkins 2006; Shirky 2008; Varnelis 2008). Rather than conceptualize everyday media engagement as "consumption" by "audiences," the term *networked publics* foregrounds the active participation of a distributed social network in the production and circulation of culture and knowledge. The growing salience of networked publics in young people's everyday lives is part of an important change in what constitutes the relevant social groups and publics that structure young people's learning and identity.

This research delves into the details of everyday youth participation in networked publics and into the ways in which parents and educators work to shape these engagements. Youths' online activity largely replicates their existing practices of hanging out and communicating with friends, but these characteristics of networked publics do create new kinds of opportunities for youth to connect, communicate, and develop their public identities. In addition to reshaping how youth participate in their given social networks of peers in school and their local communities, networked publics also open new avenues for youth participation through interest-driven networks.

Peer-Based Learning

Our attention to youth perspectives, as well as the high level of youth engagement in social and recreational activities online, determined our attention to the more informal and loosely organized contexts of peer-based learning. Our focus is on describing learning outside of school, primarily in settings of peer-based interaction. Although parents and educators often lament the influence of peers, as exemplified by the phrase *peer pressure*, we approach these informal social settings as spaces of opportunity for learning. Our cases demonstrate that some of the drivers of self-motivated learning come not from the institutionalized "authorities" in youth's lives setting standards and providing instruction but from their observing and communicating with people engaged in the same interests, and in the same struggles for status and recognition, that they are.

Both interest-driven and friendship-driven participation rely on peer-based learning dynamics, which have a different structure than formal instruction or parental guidance. Our description of friendship-driven learning describes a familiar genre of peer-based learning, in which online networks are supporting those sometimes painful but important lessons in growing up, giving youth an environment to explore romance, friendship, and status just as their predecessors did. Just like friendship-driven networks, interest-driven networks are also sites of peer-based learning, but they represent a different genre of participation, in which specialized interests are what bring a social group together. The peers whom youth are learning from in interest-driven practices are not defined by their given institution of school but rather through more intentional and chosen affiliations. In the case of youth who have become immersed in interest-driven publics, whom they identify as peers changes, as does the context for how peer-based reputation works. They also receive recognition for different forms of skill and learning.

New Media Literacy

Our work examines the current practices of youth and queries what kinds of literacies and social competencies they are defining as a particular generational cohort experimenting with a new set of media technologies. We have attempted to momentarily suspend our own value judgments about youth engagement with new media in an effort to better understand and appreciate what youth themselves see as important forms of culture, learning, and literacy. To inform current debates over the definition of new media literacy, we describe the forms of competencies, skills, and literacy practices that youth are developing through media production and online communication. Our work is in line with that of other scholars (e.g., Chávez and Soep 2005; Hull 2003; Mahiri 2004) who explore literacies in relation to ideology, power, and social practice in other settings where youth are pushing back against dominant definitions of literacy that structure their everyday life worlds.

We have identified certa   ⌐y practices that youth have been central participants   g: deliberately casual forms of online speech, nuanced   ⌐orms for how to engage in social network activities, an   ⌐enres of media representation, such as machinima, ⌐   ⌐, remix, video blogs, Web comics, and fansubs. Often   tural forms are tied to certain linguistic styles identifi   ⌐   particular youth cultures and subcultures (Eckert 1996). ⌐   ⌐al of our work is to situate these literacy practices within specific and diverse conditions of youth culture and identity as well as within an intergenerational struggle over literacy norms.

Genres of Participation with New Media

Our goal has been to arrive at a description of everyday youth new media practice that sheds light on related social practices and learning dynamics. Hanging out, messing around, and geeking out are three genres of participation that describe different forms of commitment to media engagement, and they correspond to different social and learning dynamics. In this last half of the report, we draw from the lengthier description in our book (Ito et al., forthcoming) to highlight the key features of these genres of participation, supported with illustrative examples. In our book, we provide more substantial ethnographic support for our findings, organized based on key domains of youth practice: friendship, intimacy, family, gaming, creative production, and work. Here we draw from this material in order to highlight the three genres of participation and focus specifically on the learning dynamics that we documented.

Hanging Out

For many American teenagers, coming of age is marked by a general shift from given childhood social relationships, such as families and local communities, to peer- and friendship-centered social groups. Although the nuances of these relationships vary in relation to ethnicity, class, and family dynamics (Austin and Willard 1998; Bettie 2003; Eckert 1989; Epstein 1998; Pascoe 2007; Perry 2002; Snow 1987; Thorne 1993), kids and teenagers throughout all of our studies invested a great deal of time and energy in creating and finding opportunities to "hang out." Unlike with other genres of participation (e.g., messing around and geeking out), parents and educators tend not to see the practices involved in hanging out as supporting learning. Many parents, teachers, and other adults we interviewed described young people's hanging out with their friends using new media as "a waste of time," and teenagers reported considerable restrictions and regulations on these activities at school, home, and in after-school centers. Moreover, availability of unrestricted computer and Internet access, competing responsibilities such as household chores and after-school activities (e.g., sports and music), and transportation frequently reflect the lack of priority adults place on hanging out.

In response to these regulations, most teenagers develop "work-arounds," or ways to subvert institutional, social, and technical barriers to hanging out (see Horst, Herr-Stephenson, and Robinson, forthcoming). These work-arounds and back channels are ways in which kids hang out together, even in settings that are not officially sanctioned for hanging out, such as the classroom, where talking socially to peers is explicitly frowned upon. Young people also use work-arounds and back channels as a strategy at home when they are separated from their friends and peers. Because these work-arounds and back channels take place in schools, homes, vehicles, and other spaces that structure young people's everyday lives, the teens who participated in our study had become adept at maintaining a continuous presence, or co-presence, in multiple contexts.

Once teens find a way to be together—online, offline, or both—they integrate new media within the informal hanging out practices that have characterized peer social life ever since the postwar era and the emergence of teens as a distinctive youth culture, a culture that continues to be tightly integrated with commercial popular-culture products targeted to teens. While the content and form of much of popular culture—music, fashion, film, and television—continue to change, the core practices of how youth engage with media as part of their hanging out with peers remain resilient (Cohen 1972; Corsaro 1985; Frank 1997; Gilbert 1986; Hine 1999; Snow 1987). This ready availability of multiple forms of media, in diverse contexts of everyday life, means that media content is increasingly central to everyday communication and identity construction. Ito (2008) uses the term *hypersocial* to define the process through which young people use specific media as tokens of identity, taste, and style to negotiate their sense of self in relation to their peers. While hanging out with their friends, youth develop and discuss their taste in music, their knowledge of television and movies, and their expertise in gaming. They also engage in a variety of new media practices, such as looking around online or playing games, when they are together with friends. For example, GeoGem, a 12-year-old Asian American girl in Silicon Valley, describes her time after school:

And then when I came home, I invited a friend over today and we decided to go through my clothes. My dad saw the huge mess in my room. I had to clean that up, but then we went on the computer. We went on Millsberry [Farms]. And she has her own account too. So she played on her account and I played on mine and then we got bored with that 'cause we were trying to play that game where we had to fill in the letters and make words out of the word. That was so hard. And we kept on trying to do it and we'd only get to level two and there's so many levels so we gave up. And we went in the garage and we played some Game Cube. And that was it and then her mom came and picked her up. (Horst, Silicon Valley Families)

In addition to gaming, which is pervasive in youth culture, technologies for storing, sharing, and listening to music, and watching, making, and uploading videos are now ubiquitous among youth. Teens frequently display their musical tastes and preferences on MySpace profiles and in other online venues by posting information and images related to favorite artists, clips and links to songs and videos, and song lyrics. Young people watch episodes of shows and short videos on YouTube when they are sitting around with their friends at home, at their friends' houses, in dorms, and even at after-school centers. The ability to download videos and browse sites such as YouTube means that youth can view media at times and in locations that are convenient and social, providing they have access to high-speed Internet. These practices have become part and parcel of sociability in youth culture and, in turn, central to identity formation among youth.

While acknowledging that not all practices online were necessarily positive (e.g., bullying, hate speech, and so on), we found that the facilitation of time and space to hang out reinforces informal, peer-based learning as well as the negotiation of identity. Through participation in social network sites such as MySpace, Facebook, and Bebo (among others) as well as instant and text messaging, young people are constructing new social norms and forms of media literacy in networked public culture that reflect the enhanced role of media in young people's lives. Some examples of these new forms of expression and social rules include the ability to mobilize tokens of media in socially meaningful ways, the construction of deliberately casual forms of online written communication, and the negotiation of norms of how to display friendships and romantic relationships online. The networked and public nature of these practices makes the "lessons" about social life (both the failures and successes) more consequential and persistent.

Always-On Communication   Young people use new media to build friendships and romantic relationships as well as to hang out with each other as much and as often as possible. This sense of being always on and engaged with one's peers involves a variety of practices, varying from the browsing of extended peer networks through MySpace and Facebook profiles to more intense ongoing exchanges of personal communication among close friends and romantic partners (Baron 2008). Youth use MySpace, Facebook, and IM to post status updates that can be viewed by the broader networked public of their peers—how they are faring in their relationships, their social lives, and other everyday activities. In turn, they can browse other people's updates to get a sense of the status of others without having to engage in direct communication. This kind of contact may also involve the exchange of relatively lightweight (in terms of content) text messages that share general moods, thoughts, or whereabouts. This keeps friends up-to-date with the happenings in different people's lives. Social network site profiles are key venues for representations of intimacy, providing a variety of ways to signal the intensity of a given relationship both through textual and visual representations.

Most of the direct personal communication that teens engage in through private messages, IM, and mobile phone communication involves exchange with close friends and romantic partners, rather than the broader peer group with whom they have more passive access. Teens usually have a "full-time intimate community" (Matsuda 2005) with whom they communicate in an always-on mode via mobile phones and IM. Derrick, a 16-year-old Dominican American living in Brooklyn, New York, explains the ways he moves between using new media and hanging out to Christo Sims (Rural and Urban Youth).

My homeboy usually be on his Sidekick, like somebody usually be on a Sidekick or somebody has a PSP or something like always are texting or something on AIM. A lot of people that I be with usually on AIM on their cell phones on their Nextels, on their Boost, on AIM or usually on their phone like he kept getting called, always getting called.

For Derrick and other teens like him, new media are integrated within their everyday hanging out practices. A white 10-year-old, dragon, who was part of Heather Horst and Laura Robinson's study of Neopets, also illustrates that hanging out together in a game is important when friends are spread across time and space. At the time of his interview with Horst, dragon had recently moved from the East Coast to California. While he was in the process of making friends at his new school, dragon regularly went online after school to play Runescape in the same server as his friends back East, talking with them via the game's written chat facility. In addition to playing and typing messages together, dragon and his friends also phoned each other using three-way calling, which dragon places on speakerphone. The sounds of 10-year-old boys arguing and yelling about who killed whom, why one person was slow, and reliving other aspects of the game filled the entire house, as if there were a house full of boys. New media such as social network sites, instant messaging programs, mobile phones, and gaming sites work as mediums for young people to extend, enhance, and hang out with people they already know.

Across the projects, we also saw evidence of a more intense form of co-presence, what Ito and Okabe call "tele-cocooning in the full-time intimate community," or the practice of maintaining frequent and sometimes constant (if passive) contact with close friends and/or romantic partners (Ito and Okabe 2005, 137). For example, C. J. Pascoe (Living Digital) has described the constant communication between Alice and Jesse, two 17-year-olds who have been dating for more than a year. The two individuals wake up together by logging onto MSN to talk between taking their showers and doing their hair. They then switch to conversing over their mobile phones as they travel to school, exchanging text messages throughout the school day. After school they tend to do their homework together at Jesse's house while Jesse plays a video game. When not together, they continue to talk on the phone and typically end the night on the phone or send a text message to say good night and "I love you" (see Pascoe, forthcoming-a). As becomes evident in the case of couples and close friends such as Jesse and Alice, many contemporary teens maintain multiple and constant lines of communication with their intimates over mobile phones, instant-message services, and social network sites, sharing a virtual space that is accessible by specific friends or romantic partners. In addition, and due in part to the affordances of media such as social network sites, many teens move beyond small-scale intimate friend groups to build "always-on" networked publics inhabited by their peers.

Flirting and Dating   While hanging out with friends online on social networking and gaming sites is one way youth extend their offline relationships, teens interested in romantic relationships also use new media to initiate the first stages of a relationship, what many teens refer to as "talking to" someone they have met and know through school or other settings. In this stage of the relationship, young people "talk" regularly over instant messaging and reference information found on sites such as MySpace and Facebook to verify and find out more information about the individuals, their friends, and their likes and dislikes. The asynchronous nature of these technologies allows teens to carefully compose messages that appear to be casual, a "controlled casualness." John, a white 19-year-old college freshman in Chicago, for instance, likes to flirt over IM because it is "easy to get a message across without having to phrase it perfectly" and "because I can think about things more. You can deliberate and answer however you want" (Pascoe, Living Digital). Like John, many teens say they often send texts or leave messages on social networking sites so that they can think about what they are going to say and play off their flirtatiousness if their object of affection does not seem to reciprocate their feelings. For example, youth use casual genres of online language to create studied ambiguity. From the outside sometimes these comments appear so casual that they might not be read as flirting, such as the following early "wall posts" by two Filipino teens—Missy and Dustin—who eventually dated quite seriously. After being introduced by mutual friends and communicating through IM, Missy, a Northern California 16-year-old, wrote on Dustin's MySpace wall: "hey . . . hm wut to say? iono lol/well i left you a comment... u sud feel SPECIAL haha =)."[9] Dustin, a Northern California 17-year-old, responded a day later by writing on Missy's wall: "hello there . . . umm i dont know what to say but at least i wrote something... you are so G!!!"[10] (Pascoe, Living Digital). Both of these comments can be construed as friendly or flirtatious, thus protecting both of the participants should one of the parties not be romantically drawn to the other. These particular comments took place in public venues on the participants' "walls" where others could read them, providing another layer of casualness and protection.

If a potential couple later becomes more serious, these same media are used to both announce a couple's relationship status as well as to further intensify and extend the relationship. Social network sites play an increasing role as couples become solidified and become what some call "Facebook official." At this point in a relationship, teens might indicate relationship status by ordering Friends[11] in a particular hierarchy, changing the formal statement of relationship status, giving gifts, and displaying pictures. Youth can also signal the varying intensity of intimate relationships through new media practices such as sharing passwords, adding Friends, posting bulletins, or changing headlines. In effect, the public nature and digital representations of these relationships require a fair degree of maintenance and, if the status of a relationship changes or ends, may also involve a sort of digital housecleaning that is new to the world of teen romance, but which has historical corollaries in ridding a bedroom or wallet of an ex-intimate's pictures (Pascoe, forthcoming-a). Given the persistence of new media—old profiles can always be saved, downloaded, copied, and circulated—the severing of a romantic relationship may also involve leaving, or changing, the social network sites in the interest of privacy.

For contemporary American teens, new media provide a new venue for their intimacy practices, a venue that renders these practices simultaneously more public and more private. Young people can now meet people, flirt, date, and break up outside of the earshot and eyesight of their parents and other adults while also doing these things in front of all of their online friends. The availability of networked public culture appears to be particularly important for marginalized youth, such as gay, lesbian, bisexual, or transgendered (GLBT) teens, as well as for teens who are otherwise marked as different and cannot easily find similar individuals in their local schools and communities. For such youth, online Web sites and other new media may emerge as a place for teens to meet different people. As C. J. Pascoe's work on the Living Digital project reveals, for many gay teens the Internet can become a place to explore their identities outside of the hetero-normativity of their everyday lives. As a result, dating Web sites and modes of communication between GLBT teens provide marginalized young people with greater opportunities to develop romantic relationships, with the same or similar level of autonomy experienced by their heterosexual peers. Moreover, participation in these online sites can represent an important source of social support and friendship.

Transformations in the Meaning of "Friends" and Friendship   Alongside changing the ways in which romantic relationships develop, the integration of Friends into the infrastructure of social net-work sites has resulted in transformations in the meaning of "friend" and "friendship" on an everyday level. Like the con-struction of deliberately casual online speech, development of social norms for how to display and negotiate online Friends involves new kinds of social and media literacy. These negotia-tions can be both enabling and awkward. For example, as Bob, a 19-year-old participant in Christo Sims's (Rural and Urban Youth) study, explains, becoming Friends on Facebook

sets up your relationship for the next time you meet them to have them be a bigger part of your life. . . . Suddenly they go from somebody you've met once to somebody you met once but also connected with in some weird Facebook way. And now that you've connected, you have to ac-knowledge each other more in person sometimes.

As Bob suggests, the corresponding ritual of Friending lays the groundwork for building a friendship. The practice of Friending not only acknowledges a connection, but it does so in a public manner. This sense of public-ness is further heightened through applications such as MySpace's "Top Friends," which encourages young people to identify and display their closest friends. Like declaring someone a best friend, the announce-ment of a preferred relationship also marginalizes others left out of the Top Friends spots and, in many instances, leads to conflict, or "drama," between friends. While these practices and conflicts were prevalent among teens in public spaces such as the school lunchroom or the mall, social network sites illumi-nate and intensify these tensions.

Although youth constantly negotiate and renegotiate the underlying social practices and norms for displaying friendship online, we have observed an emerging consensus about socially appropriate behavior that largely mirrors what is socially appropriate in offline contexts (boyd 2007, forthcoming). As at school, the process of adding and deleting Friends is a core element of participation on social network sites, one that is reinforced through passwords, nicknames, and other tools that facilitate the segmentation of their friend and peer worlds. Young people's decisions surrounding whom they accept and thus consider a Friend determines an individual's direct access to the content on their profile pages as well as the ways in which their decisions may affect others. These processes make social status and friendship more explicit and public, providing a broader set of contexts for observing these informal forms of social evaluation and peer-based learning. In other words, it makes peer negotiations visible in new ways, and it provides opportunities to observe and learn about social norms from peers.

Finally, and despite the perception that online media are enabling teens to reach out to a new set of social relations online, we have found that for the vast majority of teens, the relations fostered in school, summer camps, sports activities, and places of worship are by far the most dominant in how they define their peers and friendships. Even when young people are online and meet strangers, kids define social network sites, online journals, and other online spaces as friend and peer spaces. Teens consider adult participation in these spaces problematic and "creepy." Furthermore, while strangers represent one category of people with whom communication on these sites feels "creepy," parents represent a different set of issues. As a 14-year-old female named Leigh in Cedar Rapids, Iowa (danah boyd, Teen Sociality in Networked Publics), suggests, "My mom found my Xanga and she would check it every single day. I'm like, 'Uh.' I didn't like that 'cause it's invasion of privacy; I don't like people invading my privacy, so." As many teenagers such as Leigh acknowledge, most of these parental acts are motivated by the protection of kids' "well-being" rather than harassment for the sake of harassment. However, kids view these acts as a violation of trust, much like parents' coming into their bedrooms without knocking or listening in on their conversations. They also see these online invasions as "clueless," ill-informed, and lacking in basic social propriety.

Media and Mediation between Generations   While young people tend to avoid their parents and other adults while using social network sites and instant-messaging programs—spaces they identify with friends and peers—a large share of young people's engagements with new media occurs in the context of home and family life. Not surprisingly, parents, siblings, and other family members use media together while they are hanging out at home with their families. Stevens, Satwicz, and McCarthy (2007), for example, describe the settings in the home around the game console where siblings and playmates move fluidly in and out of game engagement with one another. Their findings are supported by the studies conducted by the Entertainment Software Association (2007), which states that 35 percent of American parents say they play computer and video games. Among "gamer parents," 80 percent report that they play video games with their children, and 66 percent say that playing games has brought their families closer together. In our studies of gaming, we found that video games are part of the common pool, or repertoire, of games and activities that kids and adults can engage in while enjoying time together socially (see Horst, forthcoming-a; Ito and Bittanti, forthcoming). Dan Perkel and Sarita Yardi discuss a 10-year-old in the San Francisco Bay area named Miguel who talked with them about playing Playstation with his dad and cousins (Digital Photo-Elicitation with Kids). As Miguel described:

Well, my dad, we used to play like every night . . . every Friday night, Saturday night, Sunday night, whatever . . . and he would invite my cousins to come over and stuff. We'd borrow games from my uncles. . . . They taught me how to play. Like, I used to . . . you know how when you play car games the car moves to the side and stuff? I would go like this with the control [moves arms wildly from side to side simulating holding a game controller as if he were racing]. So . . . they taught me how to keep still and look.

Although boys most closely identified with games, many of the girls we interviewed noted that they often played games such as Mario Kart, Dance Dance Revolution, and other popular games with their brothers when they were hanging out at home on the weekends or evenings. Other families engaged in ambient conversations while playing games, creating an atmosphere of sociality and communion around new media.

While gaming and television watching (using Tivo and other DVR devices) were the most pervasive shared family activities, one of the most interesting developments involved families who engaged in digital media production activities together. In these spaces, kids take advantage of the media available at home and get help from their parents with some of the more technical aspects of the devices. Among middle-class families these were often digital cameras, video cameras, and other editing software, and parents (typically fathers) often mobilized around their kids by trying to learn about and buy new things. In the case of the Miller family in Silicon Valley (Horst, Silicon Valley Families), the kids used a video camera at a family reunion and took turns helping to edit and sort through the best footage. In families such as the Millers (see Horst, forthcoming-b), parents use new media in their efforts to stay involved with, keep abreast of, and even participate in their kids' interests. Even if they were not part of the technology industry, as the Millers were, we found this level of involvement in other families with less confidence and knowledge of new media. In some cases, kids play an important role as the technology "expert" or "broker" in the family, translating Web sites and other forms of information for their parents. Twelve-year-old Michelle in Lisa Tripp and Becky Herr-Stephenson's study (Los Angeles Middle Schools) notes that she taught her mother, a single parent from El Salvador, how to use the computer, send emails, and do other activities (see Tripp, forthcoming). Michelle says that "I taught her how to, like . . . sometimes, she wants to upload pictures from my camera, and I show her, but she doesn't remember, so I have to do it myself. Mostly, I have to do the picture parts. I like doing the pictures." In contrast to the generational tensions that are so often emphasized in the popular media, families do come together around new media to share media and knowledge, play together, and stay involved in each other's lives.

Messing Around

Unlike hanging out, in which the desire is to maintain social connections to friends, messing around represents the beginning of a more intense media-centric form of engagement. When messing around, young people begin to take an interest in and focus on the workings and content of the technology and media themselves, tinkering, exploring, and extending their understanding. Some activities that we identify as messing around include looking around and searching for information online and experimentation and play with gaming and digital media production. Messing around is often a transitional genre in which kids move between hanging out and friendship-driven forms of participation to more interest-driven geeked-out ones. It involves experimentation and exploration with relatively low investment, where there are few consequences for failure, trial, and error.

Messing around with new media requires an interest-driven orientation and is supported by access to online resources, media production resources, as well as a social context for sharing of media knowledge and interests. Online and digital media provide unique supports for tinkering and self-exploration. When something piques their interest, given access to the Internet, young people can easily look around online. As Eagleton and Dobler (2007), Hargittai (2004, 2007), Robinson (2007), and others have noted, the growing availability of information in online spaces has started to transform young people's attitudes toward the availability and accessibility of information (Hargittai and Hinnant 2006; USC Center for the Digital Future 2004). Among our study participants who completed the Digital Kids Questionnaire, 87 percent ($n = 284$) reported using a search engine at least once per week, varying from Google, Yahoo!, and Wikipedia to other more specialized sites for information.[12]

The youth we spoke to who were deeply invested in specific media practices often described a period in which they discovered their own pathways to relevant information by looking around with the aid of search engines and other forms of online exploration. While the lack of local resources can make some kids feel isolated or in the dark, the increasing availability of search engines and networked publics where they can "lurk" (such as in Web forums, chat channels, and so on) effectively lowers the barriers to entry and thus makes it easier to look around and, in some cases, dabble or mess around anonymously. In addition to online information and resources, digital production tools also enable messing around in the forms of casual media creation, customization, and tinkering.

We find that messing around with new media is generally conducted in a context of social exchange involving media and technology. This social context can be the family, friendship-driven networks, interest-driven networks, or educational programs such as computer clubs and youth media centers. The most important factors are the availability of technical resources and a context that allows for a degree of freedom and autonomy in self-directed learning and exploration. In contrast to learning that is oriented toward a set, predefined goal, messing around is largely self-directed, and the outcomes of the activity emerge through exploration.

Getting Started   The youth we spoke to who were invested in specific media practices often described a period in which they first began looking around online for some area of interest and eventually discovered a broader palette or resources to experiment with, or an interest-driven online group. For example, Derrick, a 16-year-old teenager born in the Dominican Republic who lives in Brooklyn, New York, also looked to online resources for initial information about how to take apart a computer. He explains to Christo Sims (Rural and Urban Youth) how he first looked around online and did a Google image search for "video card" so he could see what it looked like. After looking at photos of where a video card is situated in a computer, he was able to install his own. He did the same with his sound card. He explains, "I learned a lot on my own that's for computers. . . . Just from searching up on Google and stuff."

In addition to searching online for information of interest, messing around can be initiated by a range of different technology-related activities. Many young people we spoke to described how they first got started messing around with digital media by capturing, modifying, and sharing personal photos and videos. Interviews with youth who are active online are often peppered with references to digital photos they have taken and shared with family and friends. Photos and videos, taken with friends and shared on sites such as PhotoBucket and MySpace, become an initial entry into digital media production. Similarly, the friendship-driven practices of setting up a MySpace profile provide an initial introduction to Web page construction. Sociable hanging out while gaming is also a pathway into messing around with technology as youth get more invested in learning the inner workings and rules underlying a particular game.

These forms of casual, personal media creation can lead to more sophisticated and engaged forms of media production. For example, Alison, an 18-year-old video creator (who is of white and Asian descent from Florida) in Sonja Baumer's study (Self-Production through YouTube) is aspiring to be a movie maker. She is also engaged in personal media creation as part of her interest in visual media.

I like watching my own videos after I've made them. I am the kind of person that likes to look back on memories, and these videos are memories for me. They show me the fun times I've had with my friends or the certain emotions I was feeling at that time. Watching my videos makes me feel happy because I like looking back on the past.

Although the practices of everyday photo and video making are familiar, the ties to digital distribution and more sophisticated forms of editing and modification open up a new set of possibilities for youth creative production. In other words, digital media help scaffold a transition from hanging out genres to messing around with more creative dimensions of photo and video creation (and vice versa).

Whether it is self-directed searching, taking personal photos and videos, or putting up a MySpace profile, what is characteristic of these initial forays into messing around is that youth are pursuing topics of personal interest. In our interviews with young people who were active digital media creators or deeply involved in other interest-driven groups, they generally described a moment when they took a personal interest in a topic and pursued it in a self-directed way (see Lange and Ito, forthcoming). This may have been catalyzed by a school project or a parent, but they eventually took this up on their own initiative. For example, one successful Web comics writer interviewed by Mizuko Ito (Anime Fans) said: "Basically, I had to self-teach myself, even though I was going to school for digital media. . . . School's more valuable for me to have . . . a time frame where I could learn on my own." Similarly, a 15-year-old white girl, Allison, from Georgia, describes how she learned to use video tools:

Trial and error, I guess. It's like any—whenever I learn anything with computers, I've taught myself how to use computers, and I consider myself very knowledgeable about them, but I just—I learn everything on my own, just figure it out, and the same with cameras. It's like a cell phone. I just figure out how to do it, and it's pretty quick and easy. (Patricia Lange, YouTube and Video Bloggers)

The media creators we interviewed often reflected this orientation by describing how they were largely self-taught, even though they might also describe the help they received from online and offline resources, peers, parents, and even teachers.

Tinkering and Exploration Messing around is an open-ended activity that involves tinkering and exploration that is only loosely goal directed. Often this can transition to more "serious" engagement in which a young person is trying to perfect a creative work or become a knowledge expert in the genre of geeking out. It is important to recognize, however, that this more exploratory mode of messing around is an important space of experimental forms of learning that open up new possibilities and engagements.

Tinkering often begins with modifying and appropriating accessible forms of media production that are widely distributed in youth culture. For example, Perkel (2008) describes the importance of copying and pasting code in the process of My-

Space profile creation, a practice in which youth will appropriate media and code from other sites to create their individual profiles. He characterizes MySpace profile creation as a process of "copy and paste" literacy, in which youth will appropriate media and code from other sites to create their individual profiles. Although this form of creative production may appear purely "derivative," young people see their profiles as expressions of their personal identities. This mode of taking up and modifying found materials has some similarities to the kinds of reframing and remixing that fan artists and fan fiction writers engage in. Some youth described how one of the main draws of MySpace was not just that this was the site that their friends were already using, but that the site seemed to allow a great deal more customization than other sites, a chance to not just socialize online, but also to display a visual identity. Ann, an 18-year-old white girl in Heather Horst's Silicon Valley Families study, saw her MySpace profile as a way to portray her personal aesthetic. She designed a MySpace page in her signature colors of pink and brown, the same colors as her bedroom.

Although young people did take time to mess around and modify their profiles, what they ended up putting online was usually not the case of planning and careful consideration, but whatever they happened to see while making or revisiting their profiles. For instance, danah boyd (Teen Sociality in Networked Publics) spoke with Shean, a 17-year-old black male from Los Angeles, who said, "I'm not a big fan of changing my background and all that. I would change mine probably every four months or three months. As long as I keep in touch with my friends or whatever, I don't really care about how it looks as long as it's, like, there." This approach toward tinkering and messing around is typical of the process through which profiles are made and modified. For youth who saw online profiles primarily as personal social spaces, this casual approach to their profiles was typical, and they tended not to update them with much frequency, or only when they grew tired of one. Nick, a 16-year-old male from Los Angeles who is of black and Native American descent, told danah boyd:

That's the main time I have fun when I'm just putting new pictures and new backgrounds on my page. I do that once every couple of months because sometimes it gets real boring. I'll be on one page. I'll log on to my profile and see the same picture every time. I'm man, I'm gonna do something new. (Teen Sociality in Networked Publics)

Similarly, we saw many instances of youth who started engaging with a new Web site or blog, or started writing a piece of fan fiction, but eventually discarded these experiments. The Internet is full of this evidence of youth experimentation in online expression.

This casual approach to messing around with media is also characteristic of a large proportion of video game play that we observed. Interactive media, because they allow for a great deal of player-level agency and customization, support messing-around activities as a regular part of game play. In the early years of gaming, the ability to do player-level modifications was minimal for most games, unless one were a game hacker and coder or it was a simulation game that was specifically designed for user authoring. Today, players take for granted the ability to modify and customize the parameters of a game. For example, we found that not only were youth constantly experimenting with the given parameters and settings of a game, but they also relied on game modifications and cheats to alter their game play. In Lisa Tripp and Becky Herr-Stephenson's study of Los Angeles immigrant families (Los Angeles Middle Schools), Herr-Stephenson had the opportunity to see how cheat codes operated in the everyday game play of Andres, a 12-year-old Mexican American. In her field notes she describes how Andres pulled out of his pocket a sheet of paper, which had game cheat codes written on it. After he uses a series of codes to "get the cops off his back," make his character invisible, and get free money, she asks him where he got the codes. He explains that he got them from some older kids. Herr-Stephenson writes: "I don't think he's ever thought about it as cheating (despite calling them 'cheat codes') and instead just thinks that such codes are a normal part of game play."

Cheat codes are an example of casual messing around with games and experimenting with their rules and boundaries. Another example of casual messing around with game parameters is players who enjoyed experimenting with the authoring tools embedded in games. Games such as Pokémon or Neopets are designed specifically to allow user authoring and customization of the player experience in the form of personal collections of customized pets (Ito 2008a; Ito and Horst 2006). This kind of customization activity is an entry point into messing around with game content and parameters. In Laura Robinson and Heather Horst's study of Neopets, one of Horst's interviewees describes the pleasures of designing and arranging homes in Neopets and Millsberry. She did not want to have to bother with playing games to accrue Neopoints to make her Neohome and instead preferred the Millsberry site, where it was easier to get money to build and customize a home:

Yeah, you get points easier and get money to buy the house easily. And I like to do interior design. And so I like to arrange my house and since they have, like, all of this natural stuff, you can make a garden. They have water and you can add water in your house [continues for a while discussing the attributes of her home].

Similarly, Emily, a 21-year-old from San Francisco, tells Matteo Bittanti (Game Play): "I played *The Sims* and built several Wii Miis. I like to personalize things, from my playlists to my games. The only problem is that after I build my characters I have no interest in playing them, and so I walk away from the game."

Whether it is the casual creation of a MySpace profile, a blog, or an online avatar, messing around involves tinkering and exploration of new spaces of possibility. Most of these activities are abandoned or only occasionally revisited in a lightweight way. While some view these activities as dead-ends or a waste of time, we see them as a necessary part of self-directed exploration in order to experiment with something that might eventually become a longer-term, abiding interest in creative

production and, in the process, youth learn computer skills they
might not have developed otherwise.

Social Contexts for Messing Around   Messing around with digital media is driven by personal interest, but it is supported by a
broader social and technical ecology. One of the primary drivers of personal media creation is sharing these media with
others. The traffic in media and practices such as profile creation is embedded within a social ecology, where the creation
and sharing of media is a friendship-driven set of practices (see
boyd, forthcoming; Pascoe, forthcoming-a). Online sites for
storing and circulating personal media are facilitating a growing set of options for sharing. Youth do not need to carry
around photo albums to share photos with their friends and
families; a MySpace profile or a camera phone will do the trick.
Consider the following observation by Dan Perkel (Judd Antin,
Christo Sims, and Dan Perkel, The Social Dynamics of Media
Production) in an after-school computer center:

Many of the kids had started to arrive early every day and would use
the computers and hang out with each other. While some kids were playing games or doing other things, Shantel and Tiffany (two apparently
African American female teenagers roughly 15 to 16 years old from a
low-income district in San Francisco) were sitting at two computers, separated by a third one between them that no one was using. They were
both on MySpace. I heard Shantel talking out loud about looking at pictures of her baby nephew on MySpace. I am fairly sure she was showing
these pictures to Tiffany. Then, she pulled out her phone and called her
sister and started talking about the pictures.

This scene that Perkel describes is an example of the role that photos archived on sites such as MySpace play in the everyday lives of youth. Shantel can pull up her photos from any Internet-connected computer to share casually with her friends, much as researchers have documented that youth do with camera phones (Okabe and Ito 2006). The fact that photos about one's life are readily available in social contexts means that visual media become more deeply embedded in the everyday communication of young people. The tinkering with MySpace profiles and the attention paid to digital photography are all part of the expectation of an audience of friends that makes the effort worthwhile. Youth look to each other's profiles, photos, videos, and online writing for examples to emulate and avoid in a peer-driven learning context that supports everyday media creation.

In the case of MySpace and other forms of media production that are widely distributed among youth, technical support is generally sought within the local friendship network. For most of the cases that we documented, at least one other person was almost always directly involved in creating kids' profiles. When asked about this, common responses were that a sibling, a cousin, or a friend showed them how to do it. In their research, Judd Antin, Christo Sims, and Dan Perkel (The Social Dynamics of Media Production) watched in one after-school program as people would call out asking for help and others would come around doing it for them (literally taking the mouse and pushing the buttons) or guiding them through the process. In an interview at a different site, Carlos, a 17-year-old Latino from the East Bay, told Perkel that he had initially found the whole profile-making process "confusing" and that he had used some free time in a Saturday program at school to ask different people to help him. Then later, when he knew what he was doing, he had shown his cousin how to add backgrounds. He says he explained to her that "you can just look around here and pick whichever you want and just tell me when you're finished and I'll get it for you."

Just as in the case of photo sharing and MySpace profiles, gamers also find support for their messing around activities in their local social relationships. Among boys, gaming has become a pervasive social activity and a context where they casually share technical and media-related knowledge. For example, several active fansubbers interviewed by Mizuko Ito in her Anime Fans study described how they initially met the members of their group through shared gaming experiences. When we had the opportunity to observe teens, particularly boys, in social settings, gaming was a frequent focus as well as topic of activity that often veered into technical subjects. In Katynka Z. Martínez's Computer Club Kids study, she notes that most of the boys associated with the club are avid gamers. After the computers in the lab became networked (in a moment they called "The Renaissance"), they would show up during lunch and even their 15-minute nutrition breaks to play *Halo* and *Counter-Strike* against one another. The hanging out with gaming was part of their participation in a technically sophisticated friendship group that focused on computer-based interests.

In other words, messing around with media is embedded in social contexts where friends and a broader peer group share a media-related interest and social focus. For most youth, they find this context in their local friendship-driven networks, grounded in popular practices such as MySpace profile creation, digital photography, and gaming. When youth transition to more focused interest-driven practices, they will generally reach beyond their local network of technical and media expertise, but the initial activities that characterize messing around are an important starting point for even these youth.

Transitions and Trajectories   Although most forms of messing around start and end with casual tinkering that does not move beyond the context of everyday peer sociability, we have observed a range of cases in which kids transitioned from messing around to the genre we describe as geeking out. We have also seen cases in which messing around has led to the eventual development of technical expertise in tinkering and fixing, which positions youth as local technology experts.

For example, 22-year-old Earendil describes the role that gaming played in his growing up and developing an interest in media technology. Earendil was largely home-schooled, and although his parents had strict limits on gaming until he and his brother were in middle school, Earendil describes how they got their "gaming kicks" at the homes of their friends with game consoles. After his parents loosened restrictions on computer time, his first social experiences online, when he was 15, were in a multiplayer game based on the novel *Ender's Game* and in online chats with fellow fans of *Myst* and *Riven*. When he started community college he fell in with "a group of local geeks, who, like myself, enjoyed playing games, etc." These experiences with online gamers and gamer friends in college provided a social context for messing around with a diverse range of media and technology, and he branched out to different interests such as game modding and video editing. He plans to eventually pursue a career in media making (Mizuko Ito, Anime Fans).

We also encountered a small number of youth who translated messing around with media to messing around with small ventures (Ito, forthcoming). Toni, a 25-year-old who emigrated from the Dominican Republic as a teen (Mizuko Ito, Anime Fans), describes how he was dependent on libraries and schools for his computer access through most of high school. This did not prevent him from becoming a technology expert, however, and he set up a small business selling *Playboy* pictures that he printed from library computers to his classmates. Zelan, a 16-year-old youth interviewed by Christo Sims (Rural and Urban Youth), first learned to mess around with digital media through video game play while his parents prospected for gold. Sims writes:

After getting immersed in the Game Boy he pursued newer and better consoles. As he did so he also learned how they worked. His parents did not like buying him gaming gear so he became resourceful. When his neighbors gave him their broken PlayStation 2, he took it apart, fixed it, and upgraded from his PlayStation 1 in the process. (Sims, forthcoming)

Driven by economic necessity, Zelan tinkered and learned how to manipulate technology. Eventually he began to market his skills as a technology fixer and now envisions the day when he will start his own business repairing computers or "just about anything computerwise." In her study of Computer Club Kids, Katynka Z. Martínez also encountered a young entrepreneur who learned the spirit of tinkering from his father, who is proficient with computers and also likes to refurbish classic Mustangs with his son. Martínez writes about Mac Man, a 17-year-old boy:

When he learned that a group of teachers were going to be throwing away their old computers, he asked if he could take them off their hands. Mac Man fixed the computers and put Windows on them. The computer club was started with these computers. Mac Man still comes to school with a small bag carrying the tools that he uses to work on computers. Teachers and other adults kept giving him computers that were broken and he had to figure out what to do with them. He fixed them and realized that he could sell them on eBay. He makes $100 profit for every computer that he sells. (Martínez, forthcoming)

These are not privileged kids who are growing up in Silicon Valley households of start-up capitalists, but rather they are working-class kids who embody the street smarts of how to hustle for money. They were able to translate their interest in tinkering and messing around into financial ventures that gave them a taste of what it might be like to pursue their own self-directed careers. While these kinds of youths are a small minority among those we encountered, they demonstrate the ways in which messing around can function as a transitional genre that leads to more sustained engagements with media and technology.

Geeking Out

For many young people, the ability to engage with media and technology in an intense, autonomous, and interest-driven way is a unique feature of the media environment of our current historical moment. Particularly for kids with newer technology and high-speed Internet at home, the Internet can provide access to an immense amount of information related to their particular interests, and it can support various forms of geeking out. This genre primarily refers to an intense commitment or engagement with media or technology, often one particular media property, genre, or type of technology. In our book, the chapters "Gaming" (Ito and Bittanti, forthcoming), "Creative Production" (Lange and Ito, forthcoming), and "Work" (Ito, forthcoming) describe some of the cases of kids who geek out on their interests and develop reputation and expertise within specialized knowledge communities. Geeking out involves learning to navigate esoteric domains of knowledge and practice and participating in communities that traffic in these forms of expertise.

Ongoing access to digital media is a requirement of intensive commitment to new media that is characteristic of geeking out. Additionally, in many of our cases, we have found that technological access is just part of what makes participation possible. Family, friends, and other peers in on- and offline spaces become particularly important in facilitating access to the technology, knowledge, and the social connections required to geek out. Just as in the case of messing around, geeking out requires the time, space, and resources to experiment and follow interests in a self-directed way. Furthermore, it requires access to specialized communities of expertise. Contrary to popular images of the socially isolated geek, almost all geeking out practices we have observed are highly social and engaged, although these are not necessarily expressed as friendship-driven social practices. Instead, geeking out is supported by specialized knowledge networks and communities that are centered on specific interests and by a range of social practices for sharing work and opinions. The online world has made these kinds of specialized hobby and knowledge networks more widely available to youth. Although generally considered marginal to both local, school-based friendship networks and academic achievement, the activities of geeking out provide important spaces of self-directed learning that are driven by passionate interests. Geeking out represents a mode of learning and developing expertise that is peer-driven but focused on gaining deep knowledge and expertise in specific areas of specialization.

Specialized Knowledge Networks   When young people geek out, they are delving into areas of interest that exceed common knowledge; this generally involves seeking expert knowledge networks outside of given friendship-driven networks. Rather than simply messing around with local friends, geeking out involves developing an identity and pride as an expert and seeking fellow experts in far-flung networks. Geeking out is usually supported by interest-based groups, either local or online, or some hybrid of the two, where fellow geeks will both produce and exchange knowledge on their subjects of interest. Rather than purely "consuming" knowledge produced by authoritative sources, geeked-out engagement involves accessing as well as producing knowledge to contribute to the knowledge network.

In her study of anime music video (AMV) creators (Anime Fans), Mizuko Ito interviewed Gepetto, an 18-year-old Brazilian fan.[13] He was first introduced to AMVs through a local friend and started messing around creating AMVs on his own. As his skills developed, however, he sought out the online community of AMV creators on animemusicvideos.org to sharpen his skills. Although he managed to interest a few of his local friends in AMV making, none of them took to it to the extent that he did. He relies heavily on the networked community of editors as sources of knowledge and expertise and for models to aspire to. In his local community he is now known as a video expert by both peers and adults. After seeing his AMV work, one of his high-school teachers asked him to teach a video workshop to younger students. He jokes that "even though I know nothing," to his local community, "I am the Greater God of video editing." In other words, the development of his identity and competence as a video editor would never have been fully supported within his local community.

In the geeked-out gaming world, players and game designers now expect that game play will be supported by an online knowledge network that provides tips, cheats, walk-throughs, mods, and reviews that are generated both by fellow players and commercial publishers. Personal knowledge exchange among local gamer friends, as well as this broader knowledge network, is a vital part of more sophisticated forms of game play that are in the geeking-out genre of engagement. While more casual players mess around by accessing cheats and hints online, more geeked-out players will consume, debate, and produce this knowledge for other players. Rachel Cody notes that the players in her study of *Final Fantasy XI* routinely used guides, produced both commercially and by fellow players. The guides assisted players in streamlining some parts of the game that otherwise took a great deal of time or resources. Cody observed that a few members of the linkshell in her study kept Microsoft Excel files with detailed notes on all their crafting in order to postulate theories on the most efficient ways of producing goods. As Wurlpin, a 26-year-old male from California, told Rachel, the guides are an essential part of playing the game. He commented, "I couldn't imagine [playing while] not knowing how to do half the things, how to go, who to talk to."

Interest-Based Communities and Organizations Interest-based geeking-out activities can be supported by a wide range of organizations and online infrastructures. Most interest groups surrounding fandom, gaming, and amateur media production are loosely aggregated through online sites such as YouTube,

LiveJournal, or DeviantArt, or more specialized sites such as animemusicvideos.org, fanfiction.net, and gaming sites such as Allakhazam or pojo.com. In addition, core participants in specific interest communities will often take a central role in organizing events and administering sites that cater to their hobbies and interests. Fan sites that cater to specific games, game guilds, or media series are proliferating on the Internet, as are specialized networks within larger sites such as LiveJournal or DeviantArt. Real-life meetings such as conventions, competitions, meet-ups, and gaming parties are also part of these kinds of distributed, player- and fan-driven forms of organization that support the ongoing life and social exchange of interest-driven groups.

As part of Mizuko Ito's case study on Anime Fans, she researched the practices of amateur subtitlers, or "fansubbers," who translate and subtitle anime and release it through Internet distribution. In our book, the chapter "Creative Production" has described some of the ways in which they form tight-knit work teams, with jobs that include translators, timers, editors, typesetters, encoders, quality checkers, and distributors (Lange and Ito, forthcoming). Fansub groups often work faster and more effectively than professional localization industries, and their work is viewed by millions of anime fans around the world. They often work on tight deadlines, and the fastest groups will turn around an episode within 24 hours of release in Japan. For this, fansubbers receive no monetary rewards, and they say that they pursue this work for the satisfaction of making anime available to fans overseas and for the pleasure they get in working with a close-knit production team that keeps in touch primarily on online chat channels and Web forums. Fansubbing is just one example of the many forms of volunteer labor and organizations that are run by fans. In addition to producing a wide range of creative works, fans also organize anime clubs, conventions, Web sites, and competitions as part of their interest-driven activities.

The issue of leadership and team organization was a topic that was central to Rachel Cody's study of *Final Fantasy XI*. Cody spent seven months observing participants in a high-level "linkshell," or guild. Although many purely social linkshells do populate *FFXI*, Cody's linkshell was an "endgame" linkshell, meaning that the group aimed to defeat the high-level monsters in the game. The linkshell was organized in a hierarchical system, with a leader and officers who had decision-making authority, and new members needed to be approved by the officers. Often the process of joining the linkshell involved a formal application and interview, and members were expected to conform to the standards of the group and perform effectively in battle as a team. The linkshell would organize "camps" where sometimes more than 150 people would wait for a high-level monster to appear and then attack with a well-planned battle strategy. Gaming can function as a site for organizing collective action, which can vary from the more lightweight arrangements of kids' getting together to play competitively to the more formal arrangements that we see in a group such as Cody's linkshell. In all of these cases, players are engaging in complex social organizations that operate under different sets of hierarchies and politics than those that occupy them in the offline world. Although these relationships are initially motivated by media-related interests, these collaborative arrangements and ongoing social exchanges often result in deep and lasting friendships with new networks of like-minded peers.

Feedback and Learning Interest-based communities that support geeking out have important learning properties that are grounded in peer-based sharing and feedback. The mechanisms for getting input on one's work and performance can vary from ongoing exchanges on online chat and forums to more formal forms of rankings, critiques, and competition. Unlike what young people experience in school, where they are graded by a teacher in a position of authority, feedback in interest-driven groups is from peers and audiences who have a personal interest in their work and opinions. Among fellow creators and community members, the context is one of peer-based reciprocity, where participants can gain status and reputation but do not hold evaluative authority over one another.

Not all creative groups we examined have a tight-knit community with established standards. YouTube, for example, functions more as an open aggregator of a wide range of video-production genres and communities, and the standards for participation and commentary differ according to the goals of particular video makers and social groups. Critique and feedback can take many forms, including posted comments on a site that displays works, private message exchanges, offers to collaborate, invitations to join other creators' social groups, and promotion from other members of an interest-oriented group. Simple five-star rating schemes, while useful in boosting ranking and visibility, were not valued as mechanisms for actually improving one's craft. Fansubbers generally thought that their audience had little understanding of what constituted a quality fansub and would take seriously only the evaluation of fellow producers. Similarly, AMV creators play down rankings and competition results based on "viewer's choice." The perception among creators is that many videos win if they use popular anime as source material, regardless of the merits of the editing. Fan fiction writers also felt that the general readership, while often providing encouragement, offered little in the way of substantive feedback.

In contrast to these attitudes toward audience feedback, a comment from a respected fellow creator carries a great deal of weight. Creators across different communities often described an inspiring moment when they received positive feedback and suggestions from a fellow creator whom they respected. In Dilan Mahendran's study (Hip-Hop Music Production), Edric, a 19-year-old Puerto Rican rapper describes his nervousness at his first recording session and the moment when he stepped out of the booth. "And everyone was like, 'Man, that was nice. I liked that.' And I was like, for real? I was like, I appreciate that. And ever since then I've just been stuck to writing, developing my style." Receiving positive feedback from peers who shared his interest in hip-hop was tremendously validating and gave him motivation to continue with his interests. Some communities have specific mechanisms for receiving informed feedback from expert peers. Animemusicvideos.org has extended reviewer forms that can be submitted for videos and it hosts a variety of competitions in which editors can enter their videos. All major anime conventions also have AMV competitions in which the best videos are selected by audiences as well as by fellow editors.

Young people participating in online writing communities can get substantive feedback from fellow writers. In fan fiction, critical feedback is provided by "beta readers," who read "fics" before they are published and give suggestions on style, plot, and grammar. Clarissa (17 years old, white), an aspiring writer and one of the participants in C. J. Pascoe's study Living Digital, participates in an online role-playing board, Faraway Lands. Aspiring members must write lengthy character descriptions to apply, which are evaluated by the site administrators. After receiving glowing reviews for her application, Clarissa has been a regular participant on the site and has developed friendships with many of the writers there. She has been doing a joint role play with another participant from Spain, and she has a friend from Oregon who critiques her work and vice versa. She explains how this feedback from fellow writers is more authentic to her than the evaluations she receives in school. "It's something I can do in my spare time, be creative and write and not have to be graded," because, "you know how in school you're creative, but you're doing it for a grade so it doesn't really count?" (Pascoe, forthcoming-b)

Recognition and Reputation In addition to providing opportunities for young people to learn and improve their craft, interest-driven groups also provide a mechanism for gaining recognition and reputation and an audience for creative work. Although audiences are not always considered the best mechanism for gaining feedback for improving their work, most participants in interest-driven communities are motivated by the fact that their work will be viewed by others or by being part of an appreciative community.

For example, zalas, a Chinese American in his early 20s and a participant in Mizuko Ito's study of Anime Fans, is an active participant in the anime fandom. zalas is an officer in his university anime club, a frequent presenter at local anime conventions, and a well-known participant in online anime forums and IRC, where he is connected to fellow fans 24/7. He will often scour the Japanese anime and game-related sites to get news that English-speaking fans do not have access to. "It's kinda like a race to see who can post the first tidbit about it." He estimates that he spends about eight hours a day online keeping up with his hobby. "I think pretty much all the time that's not school, eating, or sleeping." He is a well-respected expert in the anime scene because of this commitment to pursuing and sharing knowledge.

Specialized video communities, such as AMVs or live-action vidding,[14] will often avoid general-purpose video-sharing sites such as YouTube because they are not targeted to audiences who are well informed about their genres of media. In fact, on one of the forums dedicated to AMVs, any instance of the term *YouTube* is automatically censored. Even within these specialized groups, however, creators do seek visibility. Most major anime conventions now will include an AMV competition in which the winning works are showcased in addition to venues for fan artists to display and sell their work. The young hip-hop artists Dilan Mahendran spoke to also participated in musical competitions that gave them visibility, particularly if they went home with awards. Even fansubbers who insist that quality and respect among peers are more important than download numbers will admit that they do track the numbers. As one fansubber in Ito's study of Anime Fans described, "Deep down inside, every fansubber wants to have their work watched, and a high amount of viewers causes them some kind of joy whether they express it or not." Fansub groups generally make their "trackers," which record the number of downloads, public on their sites.

Young people can use large sites such as MySpace and YouTube as ways of disseminating their work to broader audiences. In Dilan Mahendran's Hip-Hop Music Production study, the more ambitious musicians would use a MySpace Music template as a way to develop profiles that situate them as musicians rather than a standard teen personal profile. Similarly, video makers who seek broader audiences gravitate toward YouTube as a site to gain visibility. YouTube creators monitor their play counts and comments for audience feedback. Frank, a white 15-year-old male from Ohio on YouTube, stated, "But then even when you get one good comment, that makes up for 50 mean comments, 'cause it's just the fact of knowing that someone else out there liked your videos and stuff, and it doesn't really matter about everyone else that's criticized you" (Patricia Lange, You-Tube and Video Bloggers).

In some cases, we have seen young people parlay their interests into income and even a sustained career. Max, a 14-year-old boy in Patricia Lange's YouTube and Video Bloggers study, turned into a YouTube sensation when he recorded his mother singing along to the Boyz II Men song playing in her headphones. She is unaware that people around her can hear her and have started to laugh. Max posts the video on YouTube and it attracts the attention of ABC's television show, *Good Morning America*, on which the video eventually airs. In the two years since it was posted, the video has received more than 2 million views and more than 5,000 text comments, many of them expressing support. Max's work has also attracted attention from another media company, which approached him about the possibility of buying another of his videos for an online advertisement. We also have cases of hip-hop artists who market their music, fan artists who sell their work at conventions, and youth who freelance as Web designers. Among the case studies of anime and Harry Potter fans, we have encountered examples

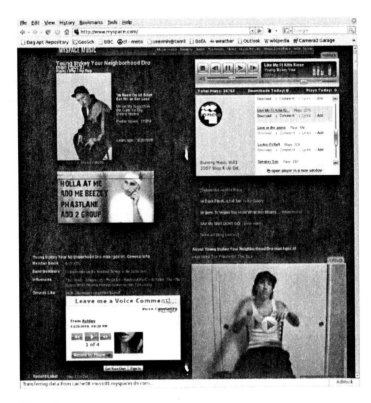

Figure 1
An Image of a MySpace Music Profile (Screen Shot by Dilan Mahendran, 2006)

of youth who have successfully capitalized on their creative talents. Becky Herr-Stephenson's study of Harry Potter fans (Harry Potter Fandom) focuses in part on podcasters who comment on the franchise. Although most podcasters are clearly hobbyists, a small number have become celebrities in the fandom who go on tours, perform "Wizard Rock," and in some cases, have gained financial rewards.

By linking "long tail" (Anderson 2006) niche audiences, online media-sharing sites make amateur- and youth-created content visible to other creators and audiences. Aspiring creators do not need to look exclusively to professional and commercial works for models of how to pursue their craft. Young people can begin by modeling more accessible and amateur forms of creative production. Even if they end there, with practices that never turn toward professionalism, they can still gain status, validation, and reputation within specific creative communities and smaller audiences. The abilities to specialize, tailor one's message and voice, and communicate with small publics are facilitated by the growing availability of diverse and niche networked publics. Gaining reputation as a rapper within the exclusive community of Bay Area Hyphy-genre hip-hop,[15] being recognized as a great character writer on a particular role-playing board, or being known as the best comedic AMV editor for a particular anime series are all examples of fame and reputation within specialized communities of interest. These aspirational trajectories do not necessarily resolve into a vision of "making it big" or becoming famous within the mode of established commercial media production. Yet they still enable young people to gain validation, recognition, and audience for their creative works and to hone their craft within groups of like-minded and expert peers. Gaining recognition in these niche and amateur groups means validation of creative work in the here and now without having to wait for rewards in a far-flung and uncertain future in creative production.

Conclusions and Implications

The goal of our project and our book has been to document the everyday lives of youth as they engage with new media and to put forth a paradigm for understanding learning and participation in contemporary networked publics. In addition to our descriptive agenda, we had a central analytic question: How do these practices change the dynamics of youth-adult negotiations over literacy, learning, and authoritative knowledge? In this concluding section, we summarize the findings of our research in relation to implications for learning, education, and public participation.

Robust participation in networked publics requires a social, cultural, and technical ecology grounded in social and recreational practices.

We have suggested that the notion of networked publics offers a framework for examining diverse forms of participation with new media in a way that is keyed to the broader social relations that structure this participation. In describing new media

engagements, we have looked at the ecology of social, technical, and cultural conditions necessary for certain forms of participation. When examining the kind of informal, peer-based
interactions that are the focus of our work, we have found that
ongoing, lightweight access to digital production tools and the
Internet was a precondition for participation in most of the
networked public spaces that are the focus of attention for U.S.
teens. Further, much of this engagement is centered on access
to social and commercial entertainment content that is generally frowned upon in formal educational settings. Sporadic,
monitored access at schools and libraries may provide sufficient
access for basic information seeking, but it is not sufficient for
the immersed kind of social engagements with networked publics that we have seen becoming a baseline for participation on
both the interest-driven and the friendship-driven sides.

Adult lack of appreciation for youth participation in popular
common cultures has created an additional barrier to access for
kids who do not have Internet access at home. We are con-
cerned about the lack of a public agenda that recognizes the
value of youth participation in social communication and popular culture. When kids lack access to the Internet at home, and
public libraries and schools block sites that are central to their
social communication, they are doubly handicapped in their
efforts to participate in common culture and sociability. These
social activities are also jumping-off points for messing around
with digital media creation and self-expression. Contemporary
social media are becoming one of the primary "institutions" of

peer culture for U.S. teens, occupying the role that was previously dominated by the informal hanging out spaces of the school, mall, home, or street. Although public institutions do not necessarily need to play a role in instructing or monitoring kids' use of social media, they can be important sites for enabling participation in these activities. Educators and policy makers need to understand that participation in the digital age means more than being able to access "serious" online information and culture; it also means the ability to participate in social and recreational activities online. This requires a cultural shift and a certain openness to experimentation and social exploration that is generally not characteristic of educational institutions.

Although we have not systematically analyzed the relation between gender and socioeconomic status and participation in interest-driven groups, our work indicates a predictable participation gap. Particularly in the cases of highly technical interest groups, and geeked-out forms of gaming, the genre itself is often defined as a masculine domain. These differences in access are not simply a matter of technology access but have to do with a more complex structure of cultural identity and social belonging. In other words, girls tend to be stigmatized more if they identify with geeked-out practices. While we may recognize that geeked-out participation has valuable learning properties, if these activities translate to downward social mobility in friendship-driven networks of status and popularity, many kids are likely to opt out even if they have the technical and social resources at their disposal. The kinds of identities and peer status that accompany certain forms of new media literacy and technical skills (and lack thereof) are areas that deserve more systematic research attention.

Networked publics provide a context for youth to develop social norms in the context of public participation.

Young people are turning to online networks to participate in a wide range of public activities. On the friendship-driven side, youth see online spaces and communications media as mechanisms to hang out with their friends. Given constraints on time and mobility, online sites offer the opportunity to casually connect with their friends as well as engage in private communication that is not monitored by parents and teachers. The ability to browse the profiles and status updates of the extended peer network in sites such as MySpace and Facebook means that youth can gain information about others in an ambient way, without engaging in direct communication. On the interest-driven side, youth turn to networked publics to connect with like-minded peers who share knowledge and expertise that may not be available to them locally. By engaging with communities of expertise online in more geeked-out practices, youth are exposed to new standards and norms for participation in specialized communities and through collaborative arrangements. These unique affordances of networked publics have altered many of the conditions of hanging out and publicity for youth, even as they build on existing youth practices of hanging out, flirting, and pursuing hobbies and interests.

In our work, contrary to fears that social norms are eroding online, we did not find many youth who were engaging in any riskier behaviors than they did in offline contexts. Youth online communication is conducted in a context of public scrutiny and structured by well-developed norms of social appropriateness, a sense of reciprocity, and collective ethics. We do not believe that educators and parents need to bear down on kids with complicated rules, restrictions, and heavy-handed norms about how they should engage online. At the same time, the actual shape of peer-based communication, and many of its outcomes, are profoundly different from those of an older generation. We found examples of parents who lacked even rudimentary knowledge of social norms for communicating online or any understanding of all but the most accessible forms of video games. Further, the ability for many youth to be in constant private contact with their peers strengthens the force of peer-based learning, and it can weaken adult participation in these peer environments. When you have a combination of a kid who is highly active online and a parent who is disengaged from these new media, we see a risk of an intergenerational wedge. Simple prohibitions, technical barriers, or time limits on use are blunt instruments; youth perceive them as raw and ill-informed exercises of power.

The problem lies not in the volume of access but in the quality of participation and learning, and kids and adults need to first be on the same page on the normative questions of learning and literacy. Parents need to begin with an appreciation of the importance of youth social interactions with their peers, an understanding of their complexities, and a recognition that children are knowledgeable experts on their own peer practices and many domains of online participation. If parents can trust that their own values are being transmitted through their ongoing communication with their kids, then new media practices can be sites of shared focus rather than sites of anxiety and tension. We believe that if our efforts to shape new media literacy are keyed to the meaningful contexts of youth participation, then there is an opportunity for productive adult engagement. Many of the norms that we observed online are very much up for negotiation, and there were often divergent perspectives among youth about what was appropriate, even within a particular genre of practice. For example, the issue of how to display social connections and hierarchies on social network sites is a source of social drama and tension, and the ongoing evolution of technical design in this space makes it a challenge for youth to develop shared social norms. Designers of these systems are central participants in defining these social norms, and their interventions are not always geared toward supporting a shared set of practices and values. More robust public debate on these issues that involves both youth and adults could potentially shape the future of online norms in this space in substantive ways.

Youth are developing new forms of media literacy that are keyed to new media and youth-centered social and cultural worlds.

In our descriptions of youth expression and online communication, we have identified a range of different practices that are evidence of youth-defined new media literacies. On the friendship-driven side, we have seen youth developing shared norms for online publicity, including how to represent oneself in online profiles, norms for displaying peer networks online, the ranking of relationships in social network sites, and the development of new genres of written communication such as composed casualness in online messages. On the interest-driven side, youth continue to test the limits on forms of new media literacy and expression. Here we see youth developing a wide range of more specialized and sometimes exclusionary forms of new media literacies that are defined in opposition to those developed in more mainstream youth practices. In geeked-out interest-driven groups, we have seen youth engage in the specialized "elite" vocabularies of gaming and esoteric fan knowledge and develop new experimental genres that make use of the authoring and editing capabilities of digital media. These include personal and amateur media that are being circulated online, such as photos, video blogs, Web comics, and podcasts, as well as derivative works such as fan fiction, fan art, mods, mashups, remixes, and fansubbing.

It is important to recognize the diverse genre conventions of youth new media literacy before developing educational programs in this space. Particularly when addressing learning and literacy that grow out of informal, peer-driven practices, we must realize that norms and standards are deeply situated in investments and identities of kids' own cultural and social worlds. For example, authoring of online profiles is an important literacy skill on both the friendship- and interest-driven sides, but one mobilizes a genre of popularity and coolness, and the other a genre of geek credibility. Similarly, the "elite" language of committed gamers involves literacies that are of little, and possibly negative, value for boys looking for a romantic partner in their school peer networks. Following from this, it is problematic to develop a standardized set of benchmarks to measure kids' levels of new media and technical literacy.

On the interest-driven side, we saw adult leadership in these groups as central to how standards for expertise and literacy are being defined. For example, the heroes of the gaming world include both teens and adults who define the identity and practice of an elite gamer. The same holds for all of the creative production groups that we examined. The leadership in this space, however, is largely cut off from the educators and policy makers who are defining standards for new media literacy in the adult-dominated world. Building more bridges between these different communities of practice could shape awareness on both the in-school and out-of-school side, if we could respond in a coordinated and mutually respectful way to the quickly evolving norms and expertise of more geeked-out and technically sophisticated experimental new media literacies.

Peer-based learning has unique properties that drive engagement in ways that differ fundamentally from formal instruction.

We see peer-based learning in networked publics in both the mainstream friendship-driven hanging out in sites such as My-Space and Facebook as well as in the more subcultural participation of geeked-out interest-driven groups. In these settings, the focus of learning and engagement is not defined by institutional accountabilities but rather emerges from kids' interests and everyday social communication. Although learning in both of these contexts is driven primarily by the peer group, the structure and the focus of the peer group differ substantially, as does the content of the learning and communication. While friendship-driven participation is largely in the mode of hanging out and negotiating issues of status and belonging in local, given peer networks, interest-driven participation happens in more distributed and specialized knowledge networks. In both the friendship-driven and interest-driven side, however, peers are an important driver of learning. Peer-based learning is characterized by a context of reciprocity, where participants feel they can both produce and evaluate knowledge and culture. Whether it is commenting on MySpace or on a fan fiction forum, participants both contribute their own content as well as comment on the content of others. More expert participants provide models and leadership but do not have authority over fellow participants. When these peer negotiations are happening in a context of public scrutiny, youth are motivated to develop their identities and reputations through these peer-based networks, exchanging comments and links and jockeying for visibility. These efforts at gaining recognition are directed at a network of respected peers rather than formal evaluations of teachers or tests. In contrast to what they experience under the guidance of parents and teachers, with peer-based learning we see youth taking on more "grown-up" roles and ownership of their own self-presentation, learning, and evaluation of others.

In contexts of peer-based learning adults can still have an important role to play, although it is not a conventionally authoritative role. In friendship-driven practices that center on sociability in given school-based networks, direct adult participation is often unwelcome, but in interest-driven groups there is a much stronger role for more experienced participants to play. Unlike instructors in formal educational settings, however, these adults participate not as educators but as passionate hobbyists and creators, and youth see them as experienced peers, not as people who have authority over them. These adults exert tremendous influence in setting communal norms and what educators might call learning goals, although they do not have direct authority over newcomers. The most successful examples we have seen of youth media programs are ones that bring kids together based on kids' own passionate interests and that have plenty of unstructured time for kids to tinker and explore without being dominated by direct instruction. Unlike classroom teachers, these lab teachers and youth-program leaders are not authority figures responsible for assessing kids' competence, but rather they are what Dilan Mahendran has called "co-conspirators," much like the adult participants in online interest-driven groups. In this, our research is in alignment with what Chávez and Soep (2005) have identified as the "pedagogy of collegiality" that defines adult-youth collaboration in what they see as successful youth media programs.

Kids' participation in networked publics suggests some new ways of thinking about the role of public education. Rather than thinking of public education as a burden that schools must shoulder on their own, what would it mean to think of public education as a responsibility of a more distributed network of people and institutions? And rather than assuming that education is primarily about preparing for jobs and careers, what would it mean to think of education as a process of guiding kids' participation in public life more generally, a public life that includes social, recreational, and civic engagement? And finally, what would it mean to enlist help in this endeavor from an engaged and diverse set of publics that are broader than what we traditionally think of as educational and civic institutions? In addition to publics that are dominated by adult interests, these publics should include those that are relevant and accessible to kids now, where they can find role models, recognition, friends, and collaborators who are coparticipants in the journey of growing up in a digital age. We hope that our research has provoked these questions.

Notes

Executive Summary

1. We use the term "social media" to refer to the set of new media that enable social interaction between participants, often through the sharing of media. Although all media are in some ways social, the term "social media" came into common usage in 2005 as a term referencing a central component of what is frequently called Web 2.0 (O'Reilly 2005 at http://www.oreillynet.com/pub/a/oreilly/tim/news/2005/09/30/what -is-web-20.html) or the social Web. All these terms refer to the layering of social interaction and online content. Popular genres of social media include: instant messaging, blogs, social network sites, and video/photo-sharing sites.

Living and Learning with New Media

1. The seven postdoctoral researchers included Sonja Baumer (University of California, Berkeley), Matteo Bittanti (University of California, Berkeley), Heather A. Horst (University of Southern California/University of California, Berkeley), Patricia G. Lange (University of Southern California), Katynka Z. Martínez (University of Southern California),

C. J. Pascoe (University of California, Berkeley), and Laura Robinson (University of Southern California).

2. The six doctoral students included danah boyd (University of California, Berkeley), Becky Herr-Stephenson (University of Southern California), Mahad Ibrahim (University of California, Berkeley), Dilan Mahendran (University of California, Berkeley), Dan Perkel (University of California, Berkeley), and Christo Sims (University of California, Berkeley).

3. The nine master's students included Judd Antin (University of California, Berkeley), Alison Billings (University of California, Berkeley), Megan Finn (University of California, Berkeley), Arthur Law (University of California, Berkeley), Annie Manion (University of Southern California), Sarai Mitnick (University of California, Berkeley), Paul Poling (University of California, Berkeley), David Schlossberg (University of California, Berkeley), and Sarita Yardi (University of California, Berkeley).

4. Judy Suwatanapongched is a JD student at the University of Southern California.

5. Rachel Cody was a project assistant at the University of Southern California.

6. The seven undergraduates are Max Besbris (University of California, Berkeley), Brendan Callum (University of Southern California), Allison Dusine (University of California, Berkeley), Lou-Anthony Limon (University of California, Berkeley), Renee Saito (University of Southern California), Tammy Zhu (University of Southern California), and Sam Jackson (Yale).

7. The collaborators include Natalie Boero, an Assistant Professor of Sociology at San Jose State University; Scott Carter, a PhD candidate at the University of California, Berkeley who now works at FXPal; Lisa Tripp, Assistant Professor of School Media and Youth Services, College

of Information, Florida State University; and Jennifer Urban, Clinical Assistant Professor of Law at the University of Southern California.

8. Full descriptions of individual research studies conducted by members of the Digital Youth project are provided online at http://digital youth.ischool.berkeley.edu/projects.

9. Like many teens, Missy wrote using typical social media shorthand. Translated, her comment would read: "Hey, hmm, what to say? I don't know. Laughing out loud. Well I left you a comment. . . . You should feel special haha (smiley face)."

10. "G" is slang for "gangsta," in this case an affectionate term for a friend.

11. We capitalize the term "Friends" when we are referring to the social network site feature for selecting Friends.

12. Although a variety of search engines are available to digital youth, across different case studies there are frequent references to Google. Some youth use various permutations such as "Googling," "Googled," and "Googler" as normative information-seeking language. The ubiquitous nature of Google may indicate that the idea of "Googling" has been normalized into the media ecology of digital youth such that for many Googling may be considered synonymous with information seeking itself.

13. Anime music videos (AMVs) are remix fan videos, in which editors will combine footage from anime with other soundtracks. Most commonly, editors will use popular Euro-American music, but some will also edit to movie trailer or TV ad soundtracks or to pieces of dialogue from movies and TV.

14. Vidding, like AMVs, is a process of remixing footage from TV shows and movies to soundtracks of an editor's choosing. Unlike AMVs, however, the live-action vidding community has been dominated by women.

15. Hyphy is a rap genre that originated in the San Francisco Bay Area and is closely associated with the late rapper Marc Dre and with Fabby Davis Junior. Hyphy music is often categorized as rhythmically up-tempo with a focus on eclectic instrumental beat arrangements, and it is also tightly coupled with particular dance styles.

References

Anderson, Chris. 2006. *The Long Tail: Why the Future of Business Is Selling Less of More*. New York: Hyperion.

Appadurai, Arjun. 1996. *Modernity at Large: Cultural Dimensions of Globalization*. Minneapolis: University of Minnesota Press.

Appadurai, Arjun, and Carol A. Breckenridge. 1988. "Why Public Culture." *Public Culture* 1, no. 1:5–9.

Austin, Joe, and Michael Nevin Willard, eds. 1998. *Generations of Youth: Youth Cultures and History in Twentieth-Century America*. New York: New York University Press.

Baron, Naomi S. 2008. *Always On: Language in an Online and Mobile World*. Oxford: Oxford University Press.

Barron, Brigid. 2006. "Interest and Self-Sustained Learning as Catalysts of Development: A Learning Ecology Perspective." *Human Development* 49, no. 4:193–224.

Basch, Linda, Nina Glick Schiller, and Christina Szanton-Blanc. 1994. *Nations Unbound: Transnational Projects, Postcolonial Predicaments, and Deterritorialized Nation-States*. London: Routledge.

Bekerman, Zvi, Nicholas C. Burbules, and Diana Silberman-Keller, eds. 2006. *Learning in Places: The Informal Education Reader.* New York: Peter Lange.

Benkler, Yochai. 2006. *The Wealth of Networks: How Social Production Transforms Markets and Freedom.* New Haven, CT: Yale University Press.

Bettie, Julie. 2003. *Women without Class: Girls, Race, and Identity.* Berkeley and Los Angeles: University of California Press.

Bourdieu, Pierre. 1984. *Distinction: A Social Critique of the Judgment of Taste.* Cambridge, MA: Harvard University Press.

boyd, danah. 2007. "Why Youth (Heart) Social Network Sites: The Role of Networked Publics in Teenage Social Life. " In *Youth, Identity, and Digital Media,* ed. D. Buckingham, 119–142. Cambridge, MA: MIT Press.

boyd, danah. Forthcoming. "Friendship." In M. Ito et al., *Hanging Out, Messing Around, and Geeking Out: Living and Learning with New Media.* Cambridge, MA: MIT Press.

Buckingham, David. 2007. *Beyond Technology: Children's Learning in the Age of Digital Culture.* Malden, MA: Polity.

Buckingham, David, ed. 2008. *Youth, Identity, and Digital Media.* Cambridge, MA: MIT Press.

Cassell, Justine, and Henry Jenkins. 1998. *From Barbie to Mortal Kombat: Gender and Computer Games.* Cambridge, MA: MIT Press.

Chávez, Vivian, and Elisabeth Soep. 2005. "Youth Radio and the Pedagogy of Collegiality." *Harvard Educational Review* 75, no. 4:409–34.

Chin, Elizabeth. 2001. *Purchasing Power: Black Kids and American Consumer Culture.* Minneapolis: University of Minnesota Press.

Cohen, Stanley. 1972. *Folk Devils and Moral Panics.* London: MacGibbon and Kee.

Corsaro, William A. 1985. *Friendship and Peer Culture in the Early Years.* Norwood, NJ: Ablex.

Corsaro, William A. 1997. *The Sociology of Childhood.* Thousand Oaks, CA: Pine Forge Press.

Eagleton, Maya B., and Elizabeth Dobler. 2007. *Reading the Web: Strategies for Internet Inquiry.* New York: Guilford.

Eckert, Penelope. 1989. *Jocks and Burnouts: Social Categories and Identity in the High School.* New York: Teachers College.

Eckert, Penelope. 1996. "Vowels and Nail Polish: The Emergence of Linguistic Style in the Preadolescent Heterosexual Marketplace." In *Gender and Belief Systems,* ed. N. Warner, J. Ahlers, L. Bilmes, M. Olver, S. Wertheim, and M. Chen, 183–190. Berkeley, CA: Berkeley Women and Language Group.

Entertainment Software Association. 2007. *Facts and Research: Top Ten Facts 2007.* http://www.theesa.com/facts/index.asp (retrieved June 28, 2007).

Epstein, Jonathon S., ed. 1998. *Youth Culture: Identity in a Postmodern World.* Malden, MA: Blackwell.

Escobar, Arturo. 1994. "Welcome to Cyberia: Notes on the Anthropology of Cyberculture." *Current Anthropology* 35, no. 3:211–31.

Fine, Gary Alan. 2004. "Adolescence as Cultural Toolkit: High School Debate and the Repertoires of Childhood and Adulthood." *The Sociological Quarterly* 24, no. 1:1–20.

Frank, Thomas. 1997. *The Conquest of Cool: Business Culture, Counterculture, and the Rise of Hip Consumerism.* Chicago: University of Chicago Press.

Gilbert, James Burkhart. 1986. *A Cycle of Outrage: America's Reaction to the Juvenile Delinquent in the 1950s.* New York: Oxford University Press.

Gilroy, Paul. 1987. *"There Ain't No Black in the Union Jack": The Cultural Politics of Race and Nation.* Chicago: University of Chicago Press.

Griffith, Maggie, and Susannah Fox. 2007. "Hobbyists Online." *Pew Internet & American Life Project.* Washington, DC: Pew.

Gupta, Akhil, and James Ferguson. 1997. *Culture Power Place: Explorations in Critical Anthropology.* Durham, NC: Duke University Press.

Hall, Stuart, and Tony Jefferson, eds. 1976. *Resistance through Rituals: Youth Subcultures in Post-War Britain.* London: Hutchinson.

Hargittai, Eszter. 2004. "Do You 'Google'? Understanding Search Engine Popularity beyond the Hype." *First Monday* 9, no. 3. http://firstmonday.org/issues/issue9_3/hargittai/index.html (retrieved June 28, 2008).

Hargittai, Eszter. 2007. "The Social, Political, Economic, and Cultural Dimensions of Search Engines: An Introduction." *Journal of Computer-Mediated Communication* 12, no. 3. http://jcmc.indiana.edu/vol12/issue3/hargittai.html (retrieved June 28, 2008).

Hargittai, Eszter, and Amanda Hinnant. 2006. "Toward a Social Framework for Information Seeking." In *New Directions in Human Information Behavior,* ed. A. Spink and C. Cole, 55–70. New York: Springer.

Hebdige, Dick. 1979. *Subculture: The Meaning of Style.* London: Routledge.

Hebdige, Dick. 1987. *Cut 'n Mix: Culture, Identity and Caribbean Music.* London: Routledge.

Hine, Thomas. 1999. *The Rise and Fall of the American Teenager.* New York: Perennial.

Holloway, Sarah L., and Gill Valentine. 2003. *Cyberkids: Children in the Information Age.* London: RoutledgeFalmer.

Horst, Heather A. Forthcoming-a. "Families." In M. Ito et al., *Hanging Out, Messing Around, and Geeking Out: Living and Learning with New Media.* Cambridge, MA: MIT Press.

Horst, Heather A. Forthcoming-b. "The Miller Family: A Portrait of a Silicon Valley Family." In M. Ito et al., *Hanging Out, Messing Around, and Geeking Out: Living and Learning with New Media.* Cambridge, MA: MIT Press.

Horst, Heather A., Becky Herr-Stephenson, and Laura Robinson. Forthcoming. "Media Ecologies." In M. Ito et al., *Hanging Out, Messing Around, and Geeking Out: Living and Learning with New Media.* Cambridge, MA: MIT Press.

Hull, Glynda. 2003. "At Last Youth Culture and Digital Media: New Literacies for New Times." *Research in the Teaching of English* 38, no. 2:229–233.

Ito, Mizuko. 2003. "Engineering Play: Children's Software and the Productions of Everyday Life." PhD diss., Anthropology, Stanford University, Palo Alto, CA.

Ito, Mizuko. 2008a. "Mobilizing the Imagination in Everyday Play: The Case of Japanese Media Mixes." In *The International Handbook of Children, Media, and Culture*, ed. K. Drotner and S. Livingstone, 397–412. Thousand Oaks, CA: Sage.

Ito, Mizuko. 2008b. "Introduction." In *Networked Publics*, ed. K. Varnelis. Cambridge, MA: MIT Press.

Ito, Mizuko. Forthcoming. "Work." In M. Ito et al., *Hanging Out, Messing Around, and Geeking Out: Living and Learning with New Media.* Cambridge, MA: MIT Press.

Ito, Mizuko, Sonja Baumer, Matteo Bittanti, danah boyd, Rachel Cody, Becky Herr-Stephenson, Heather A. Horst, Patricia G. Lange, Dilan Mahendran, Katynka Z. Martínez, C. J. Pascoe, Dan Perkel, Laura Robinson, Christo Sims, and Lisa Tripp. Forthcoming. *Hanging Out, Messing Around, and Geeking Out: Living and Learning with New Media*. Cambridge, MA: MIT Press.

Ito, Mizuko, and Matteo Bittanti. Forthcoming. "Gaming." In M. Ito et al., *Hanging Out, Messing Around, and Geeking Out: Living and Learning with New Media*. Cambridge, MA: MIT Press.

Ito, Mizuko, and Heather Horst. 2006. "Neopoints and Neo Economies: Emergent Regimes of Value in Kids Peer-to-Peer Networks." American Anthropological Association Meetings, November 16, San Jose, CA. http://www.itofisher.com/mito/itohorst.neopets.pdf (retrieved September 18, 2008).

Ito, Mizuko, and Daisuke Okabe. 2005. "Intimate Connections: Contextualizing Japanese Youth and Mobile Messaging." In *Inside the Text: Social, Cultural and Design Perspectives on SMS*, ed. R. Harper, L. Palen, and A. Taylor, 127–143. New York: Springer.

Ito, Mizuko, Daisuke Okabe, and Misa Matsuda, eds. 2005. *Personal, Portable, Pedestrian: Mobile Phones in Japanese Life*. Cambridge, MA: MIT Press.

James, Allison, and Alan Prout, eds. 1997. *Constructing and Reconstructing Childhood: Contemporary Issues in the Sociological Study of Childhood*. Philadelphia, PA: RoutledgeFarmer.

Jenkins, Henry. 1992. *Textual Poachers: Television Fans and Participatory Culture*. New York: Routledge.

Jenkins, Henry. 2006. *Convergence Culture: Where Old and New Media Collide*. New York: New York University Press.

Karaganis, Joe. 2007. "Presentation." In *Structures of Participation in Digital Culture*, ed. J. Karaganis, 8–16. New York: Social Science Research Council.

Kendall, Lori. 2002. *Hanging Out in the Virtual Pub*. Berkeley and Los Angeles: University of California Press.

Lange, Patricia G. 2008. "Terminological Obfuscation in Online Research." In *Handbook of Research on Computer Mediated Communication*, ed. S. Kelsey and K. St. Amant, 436–450. Hershey, PA: IGI Global.

Lange, Patricia G., and Mizuko Ito. Forthcoming. "Creative Production." In M. Ito et al., *Hanging Out, Messing Around, and Geeking Out: Living and Learning with New Media*. Cambridge, MA: MIT Press.

Lave, Jean, and Etienne Wenger. 1991. *Situated Learning: Legitimate Peripheral Participation*. Cambridge, UK: Cambridge University Press.

Lenhart, Amanda, Mary Madden, Alexandra Rankin Macgill, and Aaron Smith. 2007. "Teens and Social Media: The Use of Social Media Gains a Greater Foothold in Teen Life as They Embrace the Conversational Nature of Interactive Online Media." *Pew Internet & American Life Project*. Washington, DC: Pew. http://www.pewInternet.org/pdfs/PIP_Teens_Social_Media_Final.pdf (retrieved June 6, 2008).

Ling, Rich. 2004. *The Mobile Connection: The Cell Phone's Impact on Society*: San Francisco: Morgan Kaufmann.

Livingstone, Sonia. 2002. *Young People and New Media*. London: Sage.

Livingstone, Sonia. 2008. "Taking Risky Opportunities in Youthful Content Creation: Teenagers' Use of Social Networking Sites for Intimacy, Privacy, and Self-Expression." *New Media & Society* 10, no. 3:393-411.

Mahiri, Jabari, ed. 2004. *What They Don't Learn in School: Literacy in the Lives of Urban Youth*. New York: Peter Lang.

Martínez, Katynka Z. Forthcoming. "Being More Than 'Just a Banker': DIY Youth Culture and DIY Capitalism in a High-School Computer Club." In M. Ito et al., *Hanging Out, Messing Around, and Geeking Out: Living and Learning with New Media*. Cambridge, MA: MIT Press.

Matsuda, Misa. 2005. "Mobile Communication and Selective Sociality." In *Personal, Portable, Pedestrian: Mobile Phones in Japanese Life*, ed. M. Ito, D. Okabe, and M. Matsuda, 123–142. Cambridge, MA: MIT Press.

Mazzarella, Sharon, ed. 2005. *Girl Wide Web: Girls, the Internet, and the Negotiation of Identity*. New York: Peter Lang.

McRobbie, Angela, and Jenny Garber. 2000 [1978]. "Girls and Subcultures." in *Feminism and Youth Subcultures*, ed. A. McRobbie, 12–25. 2nd ed. London: Routledge.

Miller, Daniel, and Don Slater. 2000. *The Internet: An Ethnographic Approach*. New York: Berg.

Okabe, Daisuke, and Mizuko Ito. 2006. "Everyday Contexts of Camera Phone Use: Steps toward Techno-Social Ethnographic Frameworks." In *Mobile Communications in Everyday Life: Ethnographic Views, Observations, and Reflections*, ed. J. Höflich and M. Hartmann, 79–102. Berlin: Frank & Timme.

Pascoe, C. J. 2007. *"Dude, You're a Fag": Masculinity and Sexuality in High School*. Berkeley and Los Angeles: University of California Press.

Pascoe, C. J. Forthcoming-a. "Intimacy." In M. Ito et al., *Hanging Out, Messing Around, and Geeking Out: Living and Learning with New Media*. Cambridge, MA: MIT Press.

Pascoe, C. J. Forthcoming-b. "'You Have Another World to Create': Teens and Online Hangouts." In M. Ito et al., *Hanging Out, Messing Around, and Geeking Out: Living and Learning with New Media*. Cambridge, MA: MIT Press.

Perkel, Dan. 2008. "Copy and Paste Literacy? Literacy Practices in the Production of a MySpace Profile." In *Informal Learning and Digital Media: Constructions, Contexts, Consequences,* ed. K. Drotner, H. S. Jensen, and K. Schroeder, 203–224. Newcastle, UK: Cambridge Scholars Press.

Perry, Pamela. 2002. *Shades of White: White Kids and Racial Identities in High School*. Durham, NC: Duke University Press.

Rainie, Lee. 2008. "Video Sharing Websites." *Pew Internet & American Life Project*. Washington, DC: Pew.

Roberts, Donald F., Ulla G. Foehr, and Victoria Rideout. 2005. *Generation M: Media in the Lives of 8–18 Year-Olds*. Menlo Park, CA: Kaiser Family Foundation.

Robinson, Laura. 2007. "Information the Wiki Way: Cognitive Processes of Information Evaluation in Collaborative Online Venues." International Communication Association Conference, May 24–28, San Francisco, CA.

Seiter, Ellen. 1993. *Sold Separately: Parents and Children in Consumer Culture*. New Brunswick, NJ: Rutgers University Press.

Seiter, Ellen. 2005. *The Internet Playground: Children's Access, Entertainment and Mis-Education*. New York: Peter Lang.

Shirky, Clay. 2008. *Here Comes Everybody: The Power of Organizing without Organizations*. New York: Penguin.

Silverstone, Roger, and Eric Hirsch. 1992. *Consuming Technologies: Media and Information in Domestic Spaces*. London: Routledge.

Sims, Christo. Forthcoming. "Technological Prospecting in Rural Landscapes." In M. Ito et al., *Hanging Out, Messing Around, and Geeking Out: Living and Learning with New Media*. Cambridge, MA: MIT Press.

Snow, Robert. 1987. "Youth, Rock 'n' Roll and Electronic Media." *Youth & Society* 18, no. 4:326–343.

Stevens, Reed, Tom Satwicz, and Laurie McCarthy. 2007. "In-Game, In-Room, In-World: Reconnecting Video Game Play to the Rest of Kids' Lives." In *The Ecology of Games: Connecting Youth, Games, and Learning*, ed. K. Salen, 41–66. Cambridge, MA: MIT Press.

Thorne, Barrie. 1993. *Gender Play: Girls and Boys in School*. New Brunswick, NJ: Rutgers University Press.

Thorne, Barrie. 2008. "'The Chinese Girls' and 'The Pokémon Kids': Children Negotiating Differences in Urban California." In *Global Comings of Age: Youth and the Crisis of Representation*, ed. J. Cole and D. Durham. Santa Fe: School for American Research.

Tripp, Lisa. Forthcoming. "Michelle." In M. Ito et al., *Hanging Out, Messing Around, and Geeking Out: Living and Learning with New Media*. Cambridge, MA: MIT Press.

USC Center for the Digital Future. 2004. *Ten Years, Ten Trends: The Digital Future Report Surveying the Digital Future, Year Four*. Los Angeles: USC Annenberg School Center for the Digital Future. http://www.digitalcenter.org/downloads/DigitalFutureReport-Year4-2004.pdf (retrieved June 28, 2008).

Varnelis, Kazys, ed. 2008. *Networked Publics*. Cambridge, MA: MIT Press.

Wyness, Michael. 2006. *Childhood and Society: An Introduction to the Sociology of Childhood*. Basingstoke, UK: Palgrave.